Agriculture: Science and Technology

Agriculture: Science and Technology

Alastair Joseph

STATES
ACADEMIC PRESS
www.statesacademicpress.com

States Academic Press,
109 South 5th Street,
Brooklyn, NY 11249, USA

Visit us on the World Wide Web at:
www.statesacademicpress.com

ISBN: 978-1-63989-035-4 (Hardback)

Cataloging-in-Publication Data

Agriculture : science and technology / Alastair Joseph.
p. cm.
Includes bibliographical references and index.
ISBN 978-1-63989-035-4
1. Agriculture. 2. Agricultural innovations. 3. Agricultural industries--Appropriate technology.
I. Joseph, Alastair.
S3 .A37 2022
630--dc23

Table of Contents

Preface

The cultivation of plants and livestock is termed as agriculture. It primarily deals with the utilization of natural resources to produce commodities which maintain life. This includes the production of food, fiber, forest products, horticultural crops, and their related services. Agriculture involves soil, growing crops and raising livestock. It also includes the preparation of plant and animal products which are to be used by the people. There are various agricultural techniques around the world such as shifting cultivation, intercropping, monoculture, etc. Their use depends on various factors such as climate, terrain, traditions and available technology. Low-technology farming is concerned with permanent crops, which comprises food grown on land that is not replanted after each harvest. Citrus trees and coffee plants are popular examples of permanent crops. Higher-technology farming focuses on crop rotation that requires knowledge of farmable land. This book studies, analyzes and upholds the pillars of agriculture and its utmost significance in modern times. It is a valuable compilation of topics, ranging from the basic to the most complex theories and principles in the field of agriculture. Those in search of information to further their knowledge will be greatly assisted by this book.

A detailed account of the significant topics covered in this book is provided below:

Chapter 1- The multidisciplinary field of biology that deals with the production and processing of food and fibre is termed as agricultural science. Some of the topics studied under agricultural science are agricultural production, agricultural practices, agricultural systems, etc. This is an introductory chapter which will briefly introduce all these significant aspects of agricultural science.

Chapter 2- The field of study that is concerned with the application of engineering science for designing principles for agricultural purposes is known as agricultural engineering. It is a multidisciplinary field that combines various disciplines of environmental, mechanical, food science and chemical engineering to improve the efficiency of farms. All these diverse facets of agricultural engineering have been carefully analyzed in this chapter.

Chapter 3- Genetically modified plants are the plants that are engineered to create enhanced crops, new colors and to deliver vaccines. Some of the techniques that are used to create genetically modified plants are protoplast isolation, somatic cell hybridization, gene gun method, and electroporation. The topics elaborated in this chapter will help in gaining a better perspective about these techniques.

Chapter 4- Agricultural water is the water used for the production of food and fibre. It also includes water used in the watering of livestock and irrigation of crops. A few of the topics studied under agricultural water management are the role of irrigation in food production, irrigation scheduling, ditch irrigation, overhead irrigation, etc. This chapter closely examines these key concepts of agricultural water management to provide an extensive understanding of the subject.

Chapter 5- All the mechanical structures and devices used in agriculture fall under agricultural machinery. Different types of machines used in farming are manure spreader, cultivator, plough, planter, threshing machine, etc. This chapter has been carefully written to provide an easy understanding of the varied facets of these machines and technologies used in agriculture.

It gives me an immense pleasure to thank our entire team for their efforts. Finally in the end, I would like to thank my family and colleagues who have been a great source of inspiration and support.

Alastair Joseph

Chapter 1

Agricultural Science: An Introduction

The multidisciplinary field of biology that deals with the production and processing of food and fibre is termed as agricultural science. Some of the topics studied under agricultural science are agricultural production, agricultural practices, agricultural systems, etc. This is an introductory chapter which will briefly introduce all these significant aspects of agricultural science.

Agricultural science is the sciences dealing with food and fibre production and processing. They include the technologies of soil cultivation, crop cultivation and harvesting, animal production, and the processing of plant and animal products for human consumption and use.

Major Divisions

The agricultural sciences can be divided into six groups. In all fields, the general pattern of progress toward the solution of specific problems or the realization of opportunities is: (1) research to more accurately define the functional requirements to be served; (2) design and development of products, processes, and other means of better serving these requirements; and (3) extension of this information to introduce improved technologies to the agricultural industries. This has proved to be a tremendously successful approach and is being used the world over.

Soil and Water Sciences

Soil and water sciences deal with the geological generation of soil, soil and water physics and chemistry, and all other factors relevant to soil fertility. Soil science began with the formulation of the theory of humus in 1809. A generation later, Liebig introduced experimental science, including a theory of the supply of soil with mineral nutrients. In the 20th century a general theory of soil fertility developed, embracing soil cultivation, the enrichment of soil with humus and nutrients, and the preparation of soil in accordance with crop demands. Water regulation, principally drainage and irrigation, is also included.

Soil and water research have made possible the use of all classes of land in more effective ways, while the control of soil erosion and deterioration has made other advances even more striking. Because the amount of water available for plant growth is one of the major limiting factors in crop production, improved tillage and terracing practices have been devised to conserve soil moisture, and soil-management and land-use practices have been developed to increase the infiltration of snow, rain, and irrigation water, thereby reducing losses caused by runoff.

Public and private researches into chemical fertilizers and soil management have made it possible for farmers to aid nature in making specific soils more productive. Much has been learned about using crop rotation, legumes, and green manure for replenishing soil humus and nitrogen; determining and supplying the major and minor nutrient needs of crops; and managing soil under irrigation, including salt control. Techniques based on these findings have been put to use on farms to improve soil fertility and increase crop yields. Between 1940 and 1965, for example, farmers in the United States more than tripled their use of chemical fertilizers, resulting in increases of 50 to 150 percent in crop yields. Soil and water research can also help limit the amount of runoff of artificial fertilizers into nearby bodies of water by timing their application with climatic conditions and using precise amounts in controlled applications. Agrochemical runoff remains a major source of water pollution, and further research is needed to improve practices and reduce the impact of industrial agriculture on the environment.

Scientists have used many sophisticated techniques to unlock a vast storehouse of knowledge about plants. In one case, chemicals tagged with radioactive isotopes were employed to follow the processes by which plants take up soil nutrients to synthesize their fruits, grains, vegetables, nuts, flowers, and fibres.

Plant Sciences

The plant sciences include applied plant physiology, nutrition, ecology, breeding and genetics, pathology, and weed science, as well as crop management. They deal primarily with two major types of crops: (1) those that represent direct human food, such as cereals, vegetables, fruits, and nuts; and (2) those that serve as feed and forage for food, companion, laboratory, and recreational animals. Special branches of these sciences have developed to deal with each of the numerous classes of plant crops—e.g., vegetables, small fruits, citrus fruits and other tree fruits, and flowers and other ornamental plants. Other specialties concern the production of raw materials for industry—cotton, hemp, sisal, and silk—although some of these are losing economic importance in the face of competition from synthetic fibres. Branches of the plant sciences that deal with such tropical crops as coffee, tea, cocoa, bananas, coconuts, sugarcane, oil palm, and pineapples, to the contrary, promise to retain their importance.

Although scientifically based plant production came of age at the end of the 19th century, it started much earlier. Instructions on sowing dates are reported in Egypt by 2000 BCE. Throughout the centuries, numerous treatises have included recommendations on how to achieve higher and more efficient yields.

The stimulus for the development of the plant sciences did not come from botany but from agricultural chemistry, the application of which led to the development of plant physiology. Field experiments were started in Rothamsted, England, in 1834, and elsewhere in Europe soon after. Improved methods of experiment design and statistical analysis made possible the comparative study of plants and their cultivation systems.

Cultivation of plants by varieties had already led in the late 18th century to the systematic selection of cereal varieties according to predicted yield. The rediscovery at the start of the 20th century of Gregor Mendel's laws of heredity and later of ways to cause mutations led to modern plant breeding, with momentous results that included the tailoring of crop varieties for regions of climatic extremes. Agronomist Norman E. Borlaug was awarded the Nobel Prize for Peace for 1970 for the development of short-stemmed wheat, a key element in the so-called Green Revolution in developing countries.

Borlaug, Norman.

Major advances in the study of plant diseases were recorded in the 19th century, and the science of plant nutrition matured in the second quarter of the 1900s. Serious calamities resulting from the introduction of plant diseases into regions where the indigenous plants had no immunity against them, and the invasion of grapes by insects and of potatoes by late blight, stimulated research efforts. During the 20th century all diseases became objects of systematic plant pathology research. Plant pathologists search for chemicals effective against microbial diseases, weeds, and various pests and seek to adjust the biotic balance to reduce losses. That chemical residue has created some health and an environmental problem has led to further scientific activity. Biological control measures may ultimately be less harmful to the environment and more specific and effective in pest and weed management.

Other research has been undertaken because consumers want better and more diverse fruits and vegetables, including organic foods. New varieties have been developed, methods have been found to ensure that fresh and processed foods arrive at retail stores in prime condition, and grocers have learned to care for these foods so that consumers receive them in the most attractive and nutritious state. The preservation and development of historical, or heirloom, crop varieties is a growing field, as these genetic lines may have favourable traits, such as resistance to disease or drought, that can be useful for industrialized varieties or as potential alternative crops.

Animal Sciences

In modern civilizations, people rely on meat, milk, and eggs as major sources of numerous nutrients. To satisfy this demand, sheep, goats, cattle, water buffalo, swine, ducks, geese, and turkeys chickens, are produced on farms all over the world. To understand how agricultural animals convert feedstuffs into the food and other commodities consumers demand, animal scientists have undertaken broad investigations using highly sophisticated techniques. The animal sciences comprise applied animal physiology, nutrition, breeding and genetics, ecology and ethology, and livestock and poultry management. In addition, diseases of food animals are the focus of many veterinary scientists.

Animal nutrition research was well-established in several centres around the world by the turn of the 20th century, and it began to flourish during the second quarter of the 1900s. Many discoveries have been made about animal metabolism and consequent nutrient requirements; the usefulness of hundreds of feedstuffs as sources of essential amino acids, vitamins, and minerals, as well as lipids and carbohydrates; the proper balance of available nutrients in the diet; nutrient supplements and feed-processing technologies; and metabolite-partitioning and growth-promoting compounds. These fundamental findings have been applied widely since 1950, bringing about improved animal feeding. Studies of life processes in farm animals have helped in developing the optimal nutriment for each animal, and human nutrition has benefitted enormously from the knowledge that has come from these investigations.

The notion that "like begets like" was already current in biblical times. Long before the science of animal genetics developed, all species of agricultural animals were subjected to selective breeding to some extent. Modifying livestock and poultry to meet consumer demands requires the application of scientific principles to the selection of superior breeding animals and planned matings. For example, consumers have come to prefer more lean tissue and less fat in meat, and so the meat-type hog was developed in two decades of intensive selection and crossbreeding starting in the 1950s. Swine now yield more lean pork, grow faster, and require less feed to reach market weight than before. By the 1980s, a laying hen of any popular genetic strain, if managed properly, could be expected to produce more than 250 eggs annually, while special meat-producing strains of chickens gain body weight at a rate of 1:2 in ratio with feed intake.

Some of the most significant research in animal breeding has been done with dairy cattle and has established the proved sire system, in which bulls are ranked according to the performance of their offspring. The use of sires proved in this way together with artificial insemination has enabled dairy farmers to improve their herds by greatly expanding the influence of genetically superior bulls. Along with increased emphasis on performance testing, efforts have been made to predict at a young age whether an individual animal will be an efficient meat, milk, or egg producer. Such success has made for earlier culling and for herds and flocks of higher genetic merit.

Animals represent renewable agricultural resources because they reproduce, and animal scientists have studied animal reproduction assiduously since the 1930s. These investigations began in the United Kingdom but were soon joined by scientists in the United States, where the work blossomed. Basic discoveries have been put to use quickly in the animal industries. Elucidation of reproductive structures and mechanisms made it possible to refine reproductive management in the 1940s, and artificial insemination made possible the widespread use of proved sires in the 1950s. Additional basic knowledge and later technological developments made practical the control of the estrous cycle and of parturition by exogenous hormones and the serial harvesting and transplantation of embryos from donor females of high merit. The result of these changes has been an increase in the reproductive rate and efficiency of all species of farm animals.

Animal ecology and ethology are relatively young branches of the animal sciences. Around the middle of the 20th century, environmental physiologists in the United States and the United Kingdom began to study agricultural animals' relations with their environment, including temperature, air, light, and diet. Interactions among environmental temperature, diet, and the animals' genetic makeup have been characterized, and great strides have been made in improving thermal-environmental management on farms. Lighting management is now essential to profitable poultry production, and the light environment is being controlled in livestock houses as well. Since the 1970s emphasis has shifted to include the behavioral adaptability of animals to their surroundings and the effects of environmental stress on the immune status of livestock and poultry. Farmers have widely adopted intensive systems of animal production, and these systems continue to present opportunities and problems to animal scientists concerned with discovering and accommodating the environmental and ethological needs of food animals.

Animal health is essential to the efficient production of wholesome animal products. An example of the economic effect of animal-disease research conducted by veterinary scientists is the control of Marek's disease, a highly contagious disease affecting the nerves and visceral organs of chickens, which resulted in a loss of more than $200,000,000 annually to the U.S. poultry industry alone. The disease was studied for more than 30 years before it was learned that it is caused by a herpes virus. Within three years of this discovery, a vaccine was developed that reduced the frequency of Marek's disease and the resultant meat condemnations in vaccinated chickens by 90 percent and increased egg production by 4 percent. Veterinary scientists also investigate the chronic infectious diseases associated with high morbidity rates and various metabolic disorders.

Food Sciences and other Post-harvest Technologies

A group of sciences and technologies underlie the processing, storage, distribution, and marketing of agricultural commodities and by-products. Modern post-harvest technology helps provide inexpensive and various food supplies for consumers, meets the demands of a variety of industrial users, and even creates replacements for fossil fuels.

Research having particular significance to post-harvest technology includes genetic engineering techniques that increase the efficiency of various chemical and biological processes and fermentations for converting biomass to feedstock and for use in producing chemicals (including alcohols) that can replace petroleum-based products. Among the expected outcomes are the manufacture of new products from reconstituted ones and the recovery of by-products that would otherwise be considered waste.

Agricultural Production

Agricultural production is the use of cultivated plants or animals to produce products for sustaining or enhancing human life. People use a vast array of agricultural products every day—these range from the clothes we wear to the paper we write on. We decorate with flowers often produced by agriculture and run our cars in part on ethanol produced by agriculture.

Agricultural crops turned into products fall into one of four groups: foods, fuels, fibers, or raw materials. Roughly 11% of the planet's land is dedicated to crop production, and close to 26% is being used for animal pastures.

1. Food: Some examples of food products are grains and cereals. Some of the crops are turned into feed and fed to animals, which then produce dairy products like milk or are turned into food for humans or other animals. Honey and farmed fish are also some examples of food products.

2. Fuel: Agricultural products can also be used to produce fuel. Ethanol—produced from corn, sugarcane, or sorghum—is the agricultural fuel product in widest use.

3. Fiber: Fiber crops include cotton (one of the top 10 crops produced in the U.S. every year), wool, and silk. Agricultural producers also use hemp to make rope and flax for linen. It's even possible to use bamboo fiber to make cloth.

4. Raw Materials: Raw materials are the products not refined or processed for use in one of the other categories. Many of the crops grown are used to feed animals that become other agricultural products.

How does Agricultural Production Work?

Land is used to grow crops that have an intended use in one of the four categories. The crops are purchased by businesses that specialize in processing them for their expected purposes and then sold to manufacturers or distributors.

The National Organic Program (NOP)—part of the U.S. Department of Agriculture—certifies only agricultural products that fall into a specific, narrow category for use. Examples of agricultural items that the NOP can certify include such things as textiles, flowers, food, seed, plants, and feed. Items intended for other uses are not certified as organic.

The basic guideline for what constitutes an agricultural product is as follows:

- Any agricultural commodity or product, whether raw or processed, including any commodity or product derived from livestock, that is marketed in the United States for human or livestock consumption.

Businesses involved in agricultural production include:

- Breeding operations.
- Farms, including crops and certain animals.
- Greenhouses and nurseries that grow their products to be sold at retail.
- Ranches.
- Tree and sod farms (if products are sold at retail and not replanted elsewhere by the grower).

Agricultural production does not include:

- Storing or preserving raw materials before the start of the production process.
- Storing, preserving, handling, or moving finished goods.
- Storing or processing agricultural products at co-ops, grain elevators, dairies, or meatpackers.
- Raising animals for personal use.

Agricultural Production Activities

Agricultural production includes these activities:

- Agriculture: Cultivating soil, planting, raising, and harvesting crops, rearing, feeding, and managing animals.
- Aquaculture: Raising private aquatic animals (fish).
- Floriculture: Growing flowering plants.
- Horticulture: Growing fruits, vegetables, and plants.
- Maple syrup harvesting.
- Silviculture: Growing and caring for fresh trees.

Agricultural Practices

Soil Preparation

Before raising a crop, the soil in which it is to be grown is prepared by ploughing, levelling, and manuring. Ploughing is the process of loosening and digging of soil using a plough. This helps in proper aeration of the soil. After ploughing, the soil is distributed evenly and levelled in the process called levelling. The soil is then manured.

Sowing

Selection of seeds of good quality crop strains is the primary stage of sowing. After the preparation of soil, these seeds are dispersed in the field and this is called sowing. Sowing can be done manually, by hand or by using seed drilling machines. Some crops like paddy are first grown into seedlings in a small area and then transplanted to the main field.

Manuring

Crops need nutrients to grow and produce yield. Thus, the supply of nutrients at regular intervals is necessary. Manuring is the step where nutritional supplements are provided and these supplements may be natural (manure) or chemical compounds (fertilizers). Manure is the decomposition product of plant and animal wastes. Fertilizers are chemical compounds consisting of plant nutrients and are produced commercially. Apart from providing nutrients to crop, manure replenishes soil fertility as well. Other methods for soil replenishment are vermicompost, crop rotation, planting of leguminous plants.

Irrigation

Irrigation is the supply of water. Sources of water can be wells, ponds, lakes, canals, dams etc. Over irrigation may lead to waterlogging and damage the crop. This frequency and interval between successive irrigation need to be controlled.

Weeding

Weeds are unwanted plants which grow among crops. They are removed by using weedicides, by manually pulling them with hands and some are removed during soil preparation.

Harvesting

Once the crop is matured, it is cut and gathered, this process is called harvesting. Followed by harvesting, grains are separated from the chaff either by threshing, or manually in small scale (winnowing).

Storage

Grains yielded are stored in granaries or bins at godowns for later use or marketing. Therefore, methods of crop protection need to be better. In order to protect grains from pest and rodents- cleaning, drying, fumigation, etc., are done prior to storing.

Agricultural Systems

Agriculture can be regarded as a system with inputs that have physical, cultural, economic and behavioural elements. In areas where farming is less developed, physical factors are usually more important, but as human inputs increases, these physical controls become less significant. This system model can be applied to all types of farming, regardless of scale or location. It is the variations in the inputs which are responsible for the different types and patterns of agriculture around the world. The leads to classifications of agriculture in which contrasts between the different types of farming are clear.

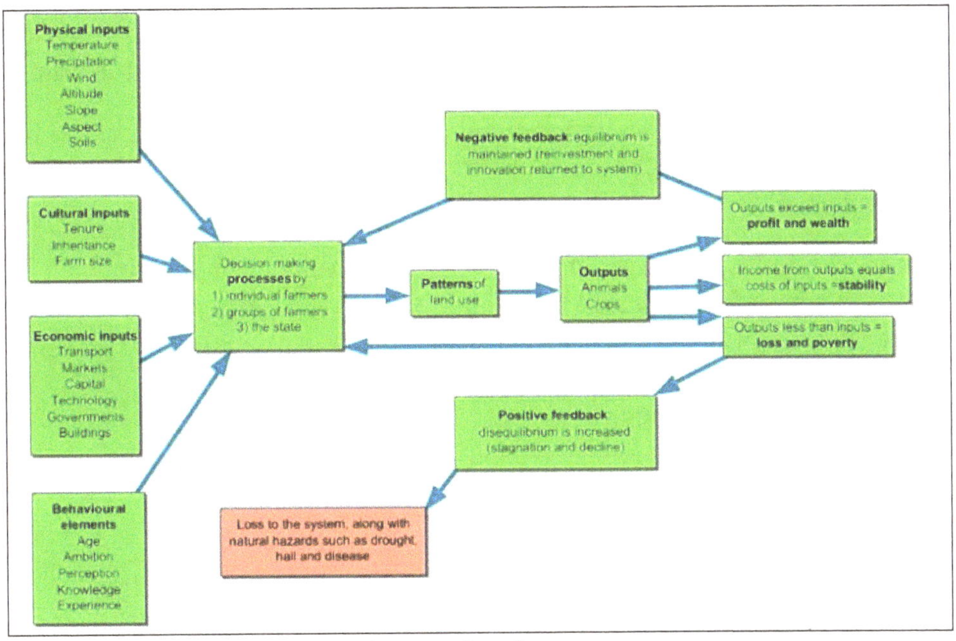

The Agricultural System.

Subsistence Agriculture

- Subsistence agriculture occurs when a plot of land produces only enough food to feed the family working it or the local community (group, tribe etc.), pay taxes and sometimes leave a little surplus for barter or to sell in better years.

- The main priority is self-sufficiency, which is achieved by growing a wide range of crops wherever possible.
- Improvements to the system are held back by a lack of capital to provide fertilisers, pesticides and other farming technology.
- Animals are kept; although where land is limited it is generally too valuable to allow grazing or growth of fodder crops.
- Where the climate is too extreme to support permanent settled agriculture, farmers become pastoral nomads, moving in search of food for their animals.
- Depending on their location, animals provide milk, meat and blood for consumption; wool and skins for shelter and clothing; dung for fuel; bones for utensils and weapons; and mounts for transport.
- Other examples of subsistence farming are shifting cultivation, which is practised in parts of the Amazon basin and in South East Asia, and wet rice agriculture, also in South East Asia and the Indian sub-continent.

Commercial Agriculture

- Commercial agriculture usually takes place on a large, profit-making scale. It may be carried out by individual farmers or by companies, with both groups trying to maximise the return on inputs and seeking maximum yields per unit of land.
- This is often achieved by growing a single crop or by raising one type of animal.
- Commercial agriculture develops in places where there are good communications and markets are large, often both domestically and on a global scale.

- Europeans have developed large-scale plantations in the tropics to supply the markets of Europe and North America with crops that include rubber, sugar cane, coffee, tea, palm oil, bananas, pineapples and tobacco.
- Other types of commercial agriculture include cattle ranching, commercial grain farming and the intensive cultivation of fruits, flowers and vegetables (sometimes referred to as market gardening).
- A growing number of farmers throughout the world are now abandoning the growth of staple food crops in order to produce for the emerging biofuels market.

Extensive and Intensive Agriculture

Extensive and intensive refer to the relationship of inputs to each other, particularly labour, capital and land. Extensive agriculture is carried out on a large scale, whereas intensive agriculture is usually relatively small scale.

- Extensive agriculture occurs when: 1) The amounts of capital and labour are small in relation to the amounts of land being farmed. Shifting cultivation is examples of farming in which labour and capital are both low but large areas are covered. 2) Labour is limited and capital higher. For example, cattle ranching and extensive grain cultivation in the USA, Canada and Australia.
- Intensive agriculture occurs when: 1) The amount of labour is high, even if the amount of capital is low in relation to the area being farmed. An example is intensive wet rice cultivation. 2) Labour input can be low but capital input high, allowing high levels of mechanisation and technology input. This occurs in intensive fruit, flower and vegetable production in the Netherlands.

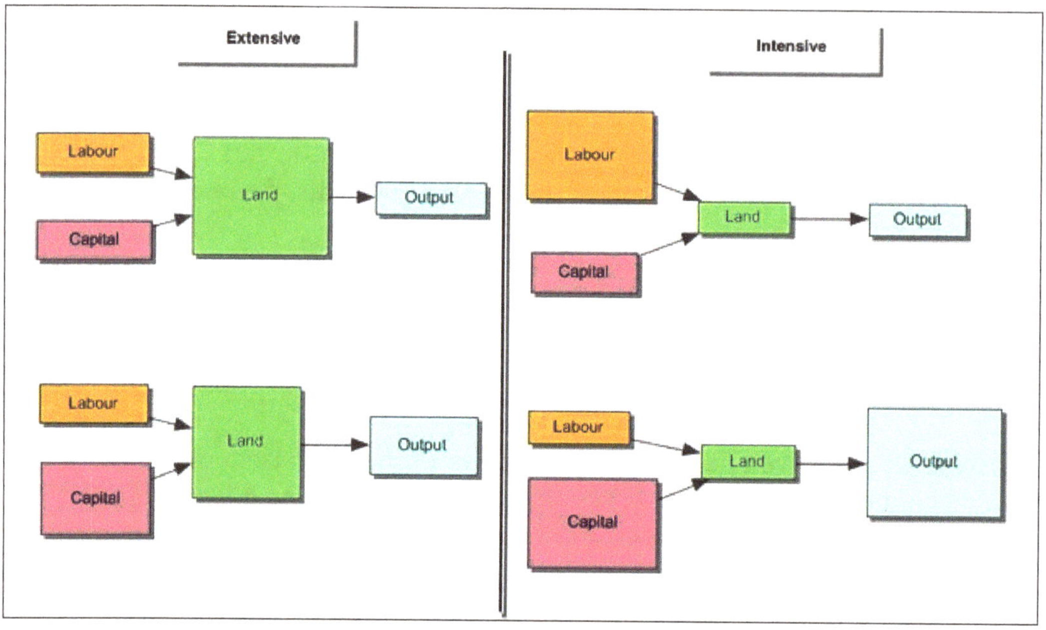

Arable, Pastoral and Mixed Farming

- Arable refers to growing crops, whether on an intensive scale (e.g. rice and market gardening) or on an extensive scale (e.g. grain farming on the Canadian prairies). Arable farming usually takes place on the more favoured land which is flatter and has high quality soil.
- Pastoral farming involves animals and usually takes place in areas less favourable to arable farming (wetter, steeper, colder, higher). This can be at an extensive scale (e.g. pastoral nomadism and cattle ranching) or intensive (e.g. dairy farming).
- Mixed farming occurs when farmers grow crops and rear animals on the same farm. It tends to take place in more developed countries where it reduces the commercial risk of relying on one type of farming. It is typical of farming in many parts of the British Isles. In subsistence areas it reduces the risk of food shortages.

Definitions of some major types of farming systems in common use by agricultural researchers, policymakers and farmers follow:

- Conservation agriculture systems employ resource-conserving methods but are also considered high-output agricultural systems. Conservation farming typically involves the integrated use of minimal tillage, cover crops and crop rotations.
- Reduced or low-input farming systems minimize the use of off-farm resources

such as commercially purchased chemicals and fuels. These systems also tighten nutrient and energy cycles and use internal resources such as biological pest controls, solar or wind energy, biologically fixed nitrogen, and other nutrients from green manures, organic matter or soil reserves. Many reduced- or low-input farming systems are examples of integrated farming systems.

- Integrated farming systems combine methods of conventional and organic production systems in an attempt to balance environmental quality and economic profit. For example, integrated farmers build their soils with composts and green manure crops but also use some synthetic fertilizers in addition to biological, cultural and mechanical pest control practices.
- Alternative livestock production systems use lower-confinement housing and rely more on pastures than do conventional and industrial livestock farms. A common example in dairy farming is the use of intensive rotational grazing practices in which short-duration; intensive grazing episodes are followed by long rest periods that allow pastures or fields to recover.
- Integrated crop–livestock farming systems generate a significant fraction of animal feed on cropland and pastures owned or managed by the livestock farmer. These systems use the crop and livestock enterprises to efficiently recycle nutrients, promote crop rotations and insulate livestock farmers from price fluctuations in feed and input markets.
- Organic agriculture is both an ecological production management system and a labeling term that indicates that the food or other agricultural product has been produced using approved methods that integrate cultural, biological and mechanical practices that foster cycling of resources, promote ecological balance and conserve biodiversity. Synthetic fertilizers, sewage sludge, irradiation and genetic engineering may not be used.
- Ecologically based farming systems emphasize the use of ecological pest management, nutrient cycling, and natural and renewable resources to enhance soil health and protect water quality. Organic and other "natural" farming systems fall under this category, relying on many common practices such as crop rotations, biological pest control, manures and avoiding all or most synthetic fertilizers and pest controls.
- Food systems refer to a complex set of activities and institutions that link food production to food consumption. Food system studies often use a "commodity chain" approach to analyze production, processing, selling and consumption.

Key Concepts of Agricultural Systems

Drawing from general systems theory and ecosystem ecology, the following concepts provide the foundation for agroecology and are essential for conceptualizing and understanding agricultural systems for research purposes. Taken together, they also provide a framework for facilitating interdisciplinary research.

Agricultural Systems are Defined by Unique Spatial and Temporal Boundaries

Agricultural system boundaries can be fixed, as is the case with a farm, for example, but systems can also be defined using subjective boundaries. In agricultural systems research, spatial and temporal boundaries are determined by research goals, the structure of the underlying environment, socioeconomic and political structures and by land-use decisions made by farmers and farm communities. For this reason, after the research questions or hypothesis is developed, the first step in delineating the system under study is to identify the physical and temporal boundaries that align with the problem being addressed. In a watershed study, for example, an experiment designed to measure the impact of spring tillage and planting on water quality could have a short-term focus. If the emphasis were to quantify nutrient loads leaving the watershed, then a multiyear study would be needed to reflect seasonal and year-to-year variability. Physical boundaries for a system vary widely, from a field or management unit or the property line for a specific farm, to a collection of farms, an entire watershed or a county.

Agricultural Systems are Composed of Interacting Subsystems

All systems are composed of many smaller, interacting subsystems that interact in either a hierarchical or nonhierarchical manner. The predominance of nested hierarchies of subsystems within agricultural and ecological systems is a striking feature. Watersheds are a prime example of a system composed of a nested hierarchy of subsystems. A large river basin, for instance, includes many smaller tributaries draining smaller watersheds, each of which has its own smaller system of tributaries and watersheds. Nested hierarchies are not always smaller versions of a larger system, as is the case with watersheds. More commonly, agricultural systems are composed of subsystems that have their own unique properties. For instance, fields may be aggregated into farms, and farms into watersheds, agricultural regions or counties. In other words, each level in the nested hierarchy is composed of smaller systems that are distinctly different from the larger system. Alternatively, subsystems can also exhibit non-hierarchical relationships as seen in an integrated farm that produces grains and animals. In this case, the two enterprises are simply interacting subsystems within the larger farming system and are connected by exchanges of crop outputs (grain and forages for animal feed) and nutrients (manure applied to fields).

System Processes occur at Different Scales and Rates

Just as system boundaries have unique physical and temporal boundaries, system processes also vary in space and time. For example, processes such as nutrient cycling occur at scales from a few microns to a whole plant and from a single field to a farming community. Similarly, time-based processes can range from minutes to centuries; decomposition of labile organic matter, or population changes in pests due to predator–prey interactions, can occur within a single growing season. In contrast, detectable changes in stabilized soil organic matter or the emergence of weed resistance to herbicides can take years or decades to manifest. During major shifts in management regimes, such as the transition from conventional to organic management or from conventional tillage to no-tillage, the rate of change for certain processes can be rapid, while other processes are not detectable for years or decades. For example, replacing fallow with cover crops can affect soil decomposers long before changes in total soil organic carbon can be detected. Because different processes will not reach dynamic steady-state conditions at the same time, the time frame of these various processes needs to be considered when planning research, particularly when focusing on the transition from one management system to another, because legacy effects from the previous system can interact with newly imposed practices.

System Structure Determines Function

In agroecosystems, structural properties (e.g., soil type, climate, biodiversity) drive functions such as plant productivity, nitrogen retention or greenhouse gas emissions, as well as emergent properties such as stability and resilience. This relationship between structure and function provides a useful framework for designing agricultural systems to optimize particular functions or for understanding the basis for differences across agroecosystems. For example, greater biodiversity in natural ecosystems often corresponds with greater productivity and enhanced resilience of the system. Thus, intentional management of species diversity can be a key strategy for achieving sustainable farming systems.

Agricultural Systems are Open Systems

Agricultural systems are open, meaning that energy, nutrients, organisms and information constantly cross system boundaries. Quantification of net flows among system components and into and out of systems, such as nutrient and energy budgets, mass balance calculations and life-cycle analysis, is important for understanding the movements and effects of these processes and properties. For example, quantification of nutrient flow across a predefined system boundary such as a field or watershed is essential to understanding the impact of farm management on long-term soil fertility and on the surrounding landscape.

Agricultural Systems have Emergent Properties

All systems have emergent properties, or characteristics and behaviors that are only apparent at higher levels of system complexity. In other words, these properties only emerge when the system is operating as a complex of subsystems; emergent properties do not exist when the subsystems or components are observed in isolation. For example, animal organ systems, such as the digestive, reproductive and cardiovascular systems, exist as such but are not viable in isolation; however, when combined in an animal structure, the emergent property of life becomes apparent. In agricultural systems, soil quality can be considered an emergent property because it exists only as a function of the interactions among soil biological, physical and chemical processes. Sustainability is also considered an emergent property, because it emerges from the multiple social and physical interactions within the system.

Agricultural Soil Science

Soils are the most important economic environment that we have for ensuring foods, raw materials, industrial developments and sustainable livelihoods of millions people. The agronomists understood that soil is the medium in which plant roots anchor and from which they extract water and nutrients. The science of soil has played and continues to plays a key role for the improvements and understanding of soil and its function in the global ecosystem for sustainable economic development. This function of soil has long been recognized in the development of various systems of agriculture, land quality, water quality and types of social, cultural and industrial civilizations. Soil science can be considered a key to sustainable crop production for economic development.

It is clear that agriculture sustains and defines our modern lives, but it is often disruptive of natural ecosystems. This is especially true for plant communities, animal populations, soil systems, and water resources. Understanding, evaluating, and balancing detrimental and beneficial agricultural disturbances of soil and water resources are essential tasks in human efforts to sustain and improve human well-being. Such knowledge influences our emerging ethics of sustainability and responsibility to human populations and ecosystems of the future.

Although agriculture is essential for human food and the stability of complex societies, almost all of our evolution has taken place in small, mobile, kin-based social groups, such as bands and tribes. Before we became sedentary people dependent on agriculture, we were largely dependent on wild plant and animal foods, without managing soil and water resources for food production. Our social evolution has accelerated since the Agricultural Revolution and taken place synergistically with human biological evolution, as we have become dependent on domesticated plants and animals grown purposefully in highly managed, soil-water systems.

Soil Fertility and Crop Growth

The early use of fire to flush out wild game and to clear forested land provided the first major anthropogenic influence on the environment. By burning native vegetation, early humans were able to gain access to herbivores grazing on the savanna and in nearby woodlands, and to suppress the growth of less desirable plant species for those easier to forage and eat. These and other factors (e.g., population pressures, climate change, encouraging/protecting desirable plants), help to lay the groundwork for the Agricultural Revolution and caused a dramatic shift in the interactions between humans and the earth. The shift from hunter-gatherer societies to an agrarian way of life drastically changed the course of human history and irreversibly altered natural nutrient cycling within soils. When humans sowed the first crop seeds at the dawn of the Neolithic Period, the soil provided plant-essential nutrients and served as the foundation for human agriculture.

Plant Nutrients

Throughout Earth's history, natural cycling of nutrients has occurred from the soil to plants and animals, and then back to the soil, primarily through decomposition of biomass. This cycling helps to maintain the essential nutrients required for plant growth in the soil. Complex nutrient cycles incorporate a range of physical, chemical, and most importantly biological processes to trace the fate of specific plant nutrients (e.g., N, P, C, S) in the environment. For a thorough analysis of these cycles, additional reference materials are available, a simplified version of nutrient cycling in natural and agricultural systems is shown in figure.

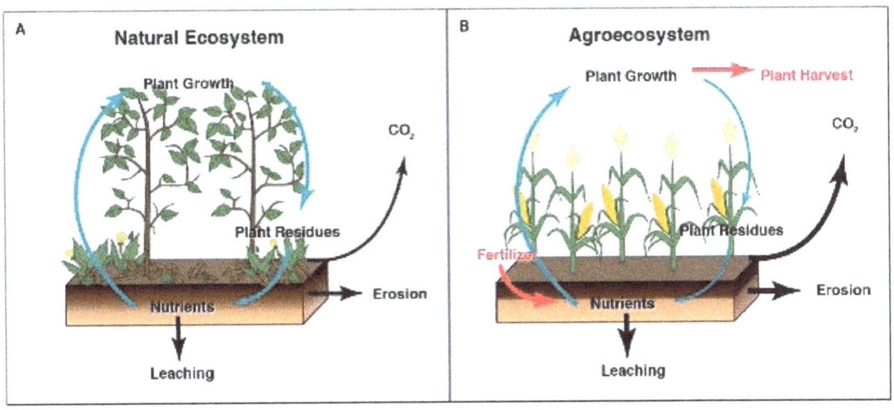

Simplified nutrient cycling schemes for (a) a natural ecosystem and (b) an agroecosystem.

Figure shows the thickness of arrows correspond to the relative amounts. The blue arrows represent basic nutrient pathways such as plant uptake, biomass decomposition, and nutrient return to soil. In the agroecosystem fertilizer (e.g., manure, compost, chemical) is applied and nutrients are removed through plant harvest (red arrows). Greater potential leaching (e.g., nitrate), erosion (e.g., soil, phosphate), and CO_2 emissions in the agroecosystem are indicated via thick black arrows.

It is generally accepted that there are 17 essential elements required for plant growth. The lack of any one of these essential nutrients, listed in table, can result in a severe limitation of crop yield — an example of the principle of limiting factors. Of the mineral elements, the primary macronutrients (N, P, and K) are needed in the greatest quantities from the soil and are the plant nutrients most likely to be in short supply in agricultural soils. Secondary macronutrients are needed in smaller quantities, are typically in sufficient quantities in soil, and therefore are not often limiting for crop growth. The micronutrients, or sometimes called trace nutrients, are needed in very small amounts and, if in excess, can be toxic to plants. Silicon (Si) and sodium (Na) are sometimes considered to be essential plant nutrients, but due to their ubiquitous presence in soils they are never in short supply.

Table: Essential plant nutrient elements and their primary form utilized by plants.

Essential plant element		Symbol	Primary form
Non-Mineral Elements			
	Carbon	C	CO_2 (g)
	Hydrogen	H	H_2O (l), H^+
	Oxygen	O	H_2O (l), O_2(g)
Mineral Elements			
Primary Macronutrients	Nitrogen	N	NH_4^+, NO_3^-
	Phosphorus	P	HPO_4^{2-}, $H_2PO_4^-$
	Potassium	K	K^+
Secondary Macronutrients	Calcium	Ca	Ca^{2+}
	Magnesium	Mg	Mg^{2+}
	Sulfur	S	SO_4^{2-}
Micronutrients	Iron	Fe	Fe^{3+}, Fe^{2+}
	Manganese	Mn	Mn^{2+}
	Zinc	Zn	Zn^{2+}
	Copper	Cu	Cu^{2+}
	Boron	B	$B(OH)_3$
	Molybdenum	Mo	MoO_4^{2-}
	Chlorine	Cl	Cl^-
	Nickel	Ni	Ni^{2+}

Agriculture alters the natural cycling of nutrients in soil. Intensive cultivation and harvesting of crops for human or animal consumption can effectively mine the soil of plant nutrients. In order to maintain soil fertility for sufficient crop yields, soil amendments are typically required. Early humans soon learned to amend their fields with animal manure, charcoal, ash, and lime ($CaCO_3$) to improve soil fertility. Today, farmers add numerous soil amendments to enhance soil fertility, including inorganic chemical fertilizers and organic sources of nutrients, such as manure or compost, often resulting in surplus quantities of primary macronutrients. The efficiency of fertilizer application and use by crops is not always optimized, and excess nutrients, especially N and P, can be transported via surface runoff or leaching from agricultural fields and pollutes surface and groundwater.

Soils for Agriculture

While soil is frequently referred to as the "fertile substrate", not all soils are suitable for growing crops. Ideal soils for agriculture are balanced in contributions from mineral components (sand: 0.05–2 mm, silt: 0.002–0.05 mm, clay: <0.002 mm), soil organic matter (SOM), air, and water. The balanced contributions of these components allow for water retention and drainage, oxygen in the root zone, nutrients to facilitate crop growth; and they provide physical support for plants. The distribution of these soil components in a particular soil is influenced by the five factors of soil formation: parent material, time, climate, organisms, and topography. Each one of these factors plays a direct and overlapping role in influencing the suitability of a soil for agriculture.

Inorganic Soil Components

As one might expect, contributions from each mineral size fraction help to provide the physical framework for a productive soil. Loamy-textured soils are commonly described as medium textured with functionally-equal contributions of sand, silt, and clay. These medium-textured soils are often considered ideal for agriculture as they are easily cultivated by farmers and can be highly productive for crop growth.

The mineral components of soil may exist as discrete particles, but are more commonly associated with one another in larger aggregates that provide structure to soil. These aggregates, or peds, play an important role in influencing the movement of water and air through soil. Sandy soils have large pore spaces and increase water drainage, but do not provide soils with many nutrients. Clay-rich soils, on the other hand, increase water holding capacity and provide many plant essential nutrients. A common measure of soil fertility is obtained by measuring the cation exchange capacity (CEC). The CEC is a measure of a soil's ability to exchange positive ions between the soil particles and solution surrounding these particles.

Due to their high surface area, clay particles can exert a large influence on various soil properties (e.g., CEC, structure, water-holding capacity), even when the percent clay content is low. Clay minerals are colloidal particles, having high surface area, with charged surfaces; permitting binding of many essential plant nutrients. The most prevalent clay-sized particles in soils fall into the class of layer-type aluminosilicates that commonly have permanent negative charge with a high CEC. Positively charged clay particles, which bind anions, include those which have pH dependent charge. The most common classes of these minerals in soils are the iron (Fe), aluminum (Al), and manganese (Mn) (hydr)oxides.

Soil Organic Matter

SOM comprises the partial or well-decomposed residues of organic biomass present in soil. SOM gives topsoil its deep black colors and rich aromas that many home gardeners and farmers of grassland soils are familiar with. Surface soils are composed of approximately 1 to 6% organic matter, with SOM decreasing with depth. The 'Great Plains' of North America and the 'Bread Basket' of Europe are some of the world's most productive agricultural soils because they developed under grassland vegetation, whose root biomass and decomposition resulted in SOM accumulation. Figure is a photograph of an organic matter-rich soil (mollisol) formed under prairie vegetation in the United States. The thick dark upper layers in this soil reflect the high SOM content. The presence of SOM is crucial for fertile soil as it provides essential plant nutrients, beneficially influences soil structure, buffers soil pH, and improves water holding capacity and aeration. The presence of organic, ionizable functional groups (e.g., carboxyl, alcoholic/phenolic OH, enol, quinone, and amine) impart charge to SOM, contributing high CEC, and pH buffering capacity.

Mollisol soil profile showing thick dark. A horizon with high organic matter content.

Soil pH

Often referred to as the master variable of soil, pH controls a wide range of physical, chemical, and biological processes and properties that affect soil fertility and plant growth. Soil pH, which reflects the acidity level in soil, significantly influences the availability of plant nutrients, microbial activity, and even the stability of soil aggregates. At low pH, essential plant macronutrients (i.e., N, P, K, Ca, Mg, and S) are less bioavailable than at higher pH values near 7, and certain micronutrients (i.e., Fe, Mn, Zn) tend to become more soluble and potentially toxic to plants at low pH values (5–6). Aluminum toxicity is also a common problem for crop growth at low pH (<5.5). Typically, soil pH values from 6 to 7.5 are optimal for plant growth; however, there are certain plants species that can tolerate — or even prefer — more acidic or basic conditions. Maintaining a narrow range in soil pH is beneficial to crop growth. SOM and clay minerals help to buffer soils to maintain a pH range optimal for plant growth. In instances where the pH is outside a desirable range, the soil pH can be altered through amendments such as lime to raise the pH. Ammonium sulfate, iron sulfate, or elemental sulfur can be added to soil to lower pH.

Soil Degradation and Crop Production

Soil forms from fresh parent material through various chemical and physical weathering processes and SOM is incorporated into soil through decomposition of plant residues and other biomass. Although these natural soil building processes regenerate the soil, the rate of soil formation is very slow. For this reason, soil should be considered a nonrenewable resource to be conserved with care for generations to come. The rate of soil formation is hard to determine and highly variable, based on the five factors of soil formation. Scientists have calculated that 0.025 to 0.125 mm of soil is produced each year from natural soil forming processes. Because of the time required to generate new soil, it is imperative that agricultural practices utilize best management practices (BMPs) to prevent soil erosion. The soil which is first eroded is typically the organic and nutrient enriched surface layer which is highly beneficial for plant growth. Thus, the primary on-site outcome is reduced crop yield as only the less fertile subsurface layers remain. Soil erosion also pollutes adjacent streams and waterways with sediment, nutrients, and agrochemicals creating serious off-site impacts.

Historically, conventional agriculture has accelerated soil erosion to rates that exceed that of soil formation. Erosion is often accelerated by agricultural practices that leave the soil without adequate plant cover and therefore exposed to raindrop splash and surface runoff or wind. Throughout human history, soil erosion has affected the ability of societies to produce an adequate food supply. Poignant examples of this can be seen in the eroded silt built up in the ancient riverbeds of Mesopotamia, making irrigation problematic, and the United States Dust Bowl of the 1930s where a devastating drought increased wind erosion, carrying fertile topsoil from the Midwest hundreds of kilometers to Washington, DC. Figure is a stunning photograph demonstrating the devastating effects of this severe wind erosion. The Dust Bowl made soil erosion a high priority in the American public consciousness of the 1930s, and it remains a top priority today.

Table: Rates of soil erosion and net formation for various lands use classes.

Measurement type	Sample size n	Mean rate of soil erosion* (mm/yr)	Net rate of soil formation† (mm/yr)
Conventional Agriculture	448	1.54 (0.32)‡	-1.52 to -1.42
Conservation Agriculture	47	0.082 (0.022)	-0.057 to +0.043
Native Vegetation	65	0.013 (0.016)	+0.012 to +0.112
Geological	925	0.029 (0.029)	-4.00 x 10^{-3} to +0.096

* Literature values.

† Net soil formation estimated using literature values for the rate of soil formation of 0.025 to 0.125 mm/yr.

‡ Parenthesis represents standard error of the mean.

Dallas, South Dakota.

Today, agricultural fields are not immune to the forces of nature (e.g., moving water, blowing wind, extremes of temperature) that caused soil erosion in the past. Figure shows the severe effects of surface runoff and soil loss in the northwestern United States. Implementation of agricultural best management practices (BMPs), and through the practice of conservation agriculture, the rate of soil loss can be reduced to approximately equal the rate of soil formation, although often still greater than those in natural systems. In addition to soil erosion, intensive land use has resulted in deforestation, water shortages, and rapidly increasing desertification of vast areas of the globe, all of which threaten the sustainability of our agricultural systems.

Damage to agricultural field.

Sustainable Soil Management

It is evident that, in order to maintain and increase food production, efforts to prevent soil degradation must become a top priority of our global society. Current population models predict a global population of between 8 and 10 billion in the next 50 years and a two-fold increase in food demand. If mismanagement of soil resources continues to diminish the fertility of the soil and the amount of productive arable land, then we will have lost a precious and essential pillar of sustainable agriculture. Sustainable agriculture is an approach to farming that focuses on production of food in a manner that can be maintained with minimal degradation of ecosystems and natural resources. This sustainable approach to agriculture strives to protect environmental resources, including soil, and provide economic profitability while maintaining social equity. The concept of sustainable agriculture is often misinterpreted to mean that chemical fertilizers and pesticides should never be used. This notion is incorrect, as sustainable agriculture should embrace those practices that provide the most beneficial services for agroecosystems and encourage long-term production of food supplies in a cultural context of the region. It cannot be overstressed that sustainable practices should not only consider crop production and profit, but must include land management strategies that reduce soil erosion and protect water resources. By embracing certain modern-day technologies, proven BMPs, and learning from the past, our society will be able to continue to conserve soil resources and produce food supplies sufficient to meet current and future population demands.

Agricultural Economics

Agricultural economics is the study of the allocation, distribution, and utilization of the resources used, along with the commodities produced, by farming. Agricultural economics plays a role in the economics of development, for a continuous level of farm surplus is one of the wellsprings of technological and commercial growth.

In general, one can say that when a large fraction of a country's population depends on agriculture for its livelihood, average incomes are low. That does not mean that a country is poor because most of its population is engaged in agriculture; it is closer to the truth to say that because a country is poor, most of its people must rely upon agriculture for a living.

Agriculture and Economic Development

As a country develops economically, the relative importance of agriculture declines. The primary reason for that was shown by the 19th-century German statistician Ernst Engel, who discovered that as incomes increase, the proportion of income spent on food declines. For example, if a family's income were to increase by 100 percent, the amount it would spend on food might increase by 60 percent; if formerly its expenditures on food had been 50 percent of its budget, after the increase they would amount to only 40 percent of its budget. It follows that as incomes increase, a smaller fraction of the total resources of society is required to produce the amount of food demanded by the population.

Progress in Farming

That fact would have surprised most economists of the early 19th century, who feared that the limited supply of land in the populated areas of Europe would determine the continent's ability to feed its growing population. Their fear was based on the so-called law of diminishing returns: that under given conditions an increase in the amount of labour and capital applied to a fixed amount of land results in a less-than-proportional increase in the output of food. That principle is a valid one, but what the classical economists could not foresee was the extent to which the state of the arts and the methods of production would change. Some of the changes occurred in agriculture; others occurred in other sectors of the economy but had a major effect on the supply of food.

In looking back upon the history of the more developed countries, one can see that agriculture has played an important part in the process of their enrichment. For one thing, if development is to occur, agriculture must be able to produce a surplus of food to maintain the growing non-agricultural labour force. Since food is more essential for life than are the services provided by merchants or bankers or factories, an economy cannot shift to such activities unless food is available for barter or sale in sufficient quantities to support those engaged in them. Unless food can be obtained through international trade, a country does not normally develop industrially until its farm areas can supply its towns with food in exchange for the products of their factories.

Economic development also requires a growing labour force. In an agricultural country most of the workers needed must come from the rural population. Thus agriculture must not only supply a surplus of food for the towns, but it must also be able to produce the increased amount of food with a relatively smaller labour force. It may do so by substituting animal power for human power or by gradually introducing labour-saving machinery.

Agriculture may also be a source of the capital needed for industrial development to the extent that it provides a surplus that may be converted into the funds needed to purchase industrial equipment or to build roads and provide public services.

Spinach field with irrigation system.

For those reasons, a country seeking to develop its economy may be well advised to give a significant priority to agriculture. Experience in the developing countries has shown that agriculture can be made much more productive with the proper investment in irrigation systems, research, fertilizers, insecticides, and herbicides.

Fortunately, many advances in applied science do not require massive amounts of capital, although it may be necessary to expand marketing and transportation facilities so that farm output can be brought to the entire population.

One difficulty in giving priority to agriculture is that most of the increase in farm output and most of the income gains are concentrated in certain regions rather than extending throughout the country. The remaining farmers are not able to produce more and actually suffer a disadvantage as farm prices decline. There is no easy answer to that problem, but developing countries need to be aware of it; economic progress is consistent with lingering backwardness, as has been seen in parts of southern Italy or in the Appalachian area of the United States.

Peasant Agriculture

One characteristic of undeveloped peasant agriculture is its self-sufficiency. Farm families in those circumstances consume a substantial part of what they produce. While some of their output may be sold in the market, their total production is generally not much larger than what is needed for the maintenance of the family. Not only is productivity per worker low under those conditions, but yields per unit of land are also low. Even where the land was originally fertile, the fertility is likely to have been depleted by decades of continuous cropping. The available manures are not sufficient, and the farmers cannot afford to purchase them elsewhere.

Rice planting.

Peasant agriculture is often said to be characterized by inertia. The peasant farmer is likely to be illiterate, suspicious of outsiders, and reluctant to try new methods; food patterns remain unchanged for decades or even centuries. Evidence, however, suggests that the apparent inertia may be simply the result of a lack of alternatives. If there is nothing better to change to, there is little point in changing. Moreover, the self-sufficient farmer is bound to want to minimize risks; since a crop failure can mean starvation in many parts of the world, farmers have been reluctant to adopt new methods if doing so would expose them to greater risks of failure.

The increased use worldwide of high-yielding varieties of rice and wheat from the 1960s showed that farmers were willing and able to adopt new crops and farming methods when their superiority was demonstrated. Those high-yielding varieties, however, required increased outlays for fertilizer, as well as expanded facilities for storage and distribution, and many developing countries were unable to afford such expenditures.

The Labour Force

As economic development proceeds, a large proportion of the farm labour force must shift from agriculture into other pursuits. That fundamental shift in the labour force is made possible, of course, by an enormous increase in output per worker as agriculture becomes modernized. That increase in output stems from various factors. Where land is plentiful, the output per worker is likely to be higher because it is possible to employ more fertilizer and machinery per worker.

Land, Output and Yields

Only a small fraction of the world's land area—about one-tenth—may be considered arable, if arable land is defined as land planted to crops. Less than one-fourth of the world's land area is in permanent meadows and pastures. The remainder is either in forests or is not being used for agricultural purposes.

There are great differences in the amount of arable land per person in the various regions of the world. The greatest amount of arable land per capita is in Oceania; the least is in China. No direct relationship exists between the amount of arable land per capita and the level of income.

The relationship between land, population, and farm production is a complex one. In traditional agriculture, where methods of production have changed little over a long period of time, production is largely determined by the quality and quantity of land available and the number of people working on the land. Until the early years of the 20th century, most of the world's increase in crop production came either from an increase in land under cultivation or from an increase in the amount of labour used per unit of land. That generally involved a shift to crops that would yield more per unit of land and required more labour for their cultivation. Wheat, rye, and millet require less labour per unit of land and per unit of food output than do rice, potatoes, or corn (maize), but generally the latter yield more food per unit of land. Thus, as population density increased, the latter groups of crops tended to be substituted for the former. That did not hold true in Europe, where wheat, rye, and millet expanded at the expense of pasture land, but those crops yielded more food per acre than did the livestock that they displaced.

Farm in Saskatchewan.

As agriculture becomes modernized, its dependence upon land as well as upon human labour decreases. Animal power and machinery are substituted for human labour; mechanical power then replaces animal power. The substitution of mechanical power for animal power reduces the need for land. The increased use of fertilizer as modernization occurs also acts as a substitute for both land and labour; the same is true of herbicides and insecticides. By making it possible to produce more per unit of land and per hour of work, less land and labour are required for a given amount of output.

Efforts to Control Prices and Production

In the second half of the 20th century, governments undertook to control both prices and output in the agricultural sector, largely in response to the pressures of the farmers themselves. In the absence of such control, farm prices tend to fluctuate more than do most other prices and the incomes of farmers fluctuate to an even greater degree. Not only are incomes in agriculture unstable, but they also tend to be lower than incomes in other economic sectors.

Instability of Prices

The instability of farm prices results from several factors. One is the relative slowness with which farmers are able to respond to changes in the demand for their product. Farmers generally must produce on the basis of expectations, and if their expectations turn out to be wrong, the resulting surplus or shortage cannot be corrected until the beginning of the next production cycle. Once a crop is planted, very little can be done to increase or decrease production in response to market prices. As long as prices cover current operating costs, such as the cost of harvesting, it pays farmers to carry through their production plans even if prices fall to a very low level. It is not unusual for the prices of particular farm products to vary by a third or a half from year to year. That extreme variability results from the relatively low responsiveness of demand to changes in price—i.e., from the fact that in order to increase sales by 5 percent it may be necessary to reduce the price by 15 percent.

Instability of Income

The instability of farm prices is accompanied by instability of farm income. While gross income from agriculture generally does not vary as much as do individual farm prices, net income may vary more than prices. In modern agriculture, costs tend to be relatively stable; the farmer is unable to compensate for a drop in prices by reducing his payments for machinery, fertilizer, or labour.

The incomes of farm workers are generally below those of other workers. There are two major reasons for that inequity. One is that in most economies the need for farm labour is declining, and each year large numbers of farm people, especially young ones, must leave their homes to seek jobs elsewhere. The difference in returns to labour is required to bring about that transfer of workers out of farming; if the transfer did not occur, farm incomes would be even more depressed. The second major reason for the income differences is that farm people generally have less education than do nonfarm people and are able to earn less at nonfarm jobs. The difference in education is of long standing and is found in all countries, developed and undeveloped; it also exists whether the national education system is highly decentralized, as in the United States, or highly centralized, as in France.

Government Intervention

Governments have employed various measures to maintain farm prices and incomes above what the market would otherwise have yielded. They have included tariffs or import levies, import quotas, export subsidies, direct payments to farmers, and limitations on production. Tariffs and import quotas can be effective only if a country normally imports some of its supply. Export subsidies result in higher prices to domestic consumers than to foreign purchasers; their use requires control over imports to prevent foreign supplies from entering the domestic market and bringing prices down. Direct payments to farmers have been used to maintain prices to consumers at reasonable levels, while assuring farmers a return above world-market levels. Limitations on production, intended to reduce supply and thus increase prices, have been used in Brazil (for coffee) and in the United States (for major crops).

Accomplishments

The effects of price and income policies are difficult to assess. The policies have unquestionably worked to raise agricultural production in the countries where they have been applied, but their usefulness as a means of enhancing the economic well-being of farm people is debatable. The governments of the industrial countries have been able to raise the returns from agriculture above the levels that would have prevailed in the absence of such intervention. In addition to maintaining prices, they provide subsidies for agricultural inputs such as tractor fuel and chemical fertilizers; they also gave assistance in consolidating small farms into larger ones and in improving farm buildings.

The level of income and the economic well-being of farm people in general are determined by many factors, including not only the prices they receive for their output but also the rate at which the economy in general is growing, the ease with which people can move from farm to nonfarm jobs, the prices they must pay for their productive inputs, and their level of education. With respect to average income per person, as distinguished from total income, the prices received and paid are probably less important than the other factors mentioned. That becomes obvious when one compares farm incomes in developed countries with those in less-developed ones; the differences in real income have to do mainly with the levels of economic development and not with farm prices or subsidies. Government efforts to increase farm prices are likely to be offset, in the long run, by an increase in the number of persons engaged in farming, and that tends to keep the returns to farm labour from rising much faster than they would in the absence of such policies.

There are two other reasons for believing that the income effects of higher farm prices or subsidies are relatively insignificant in the long run compared with other factors affecting incomes of farm workers. One is that an increase in farm prices induces farmers to use more fertilizer, machinery, fuel and oil, and other items. If a significant part of any increase in gross income is used for such things, the absolute increase in net farm income is much smaller than the increase in gross farm income. The second reason is that a given increase in government-supported farm prices generally occurs only once. After the increase in returns has been realized, the higher farm prices contribute nothing further to incomes. In contrast, general economic growth along with the continued reduction of the farm labour force has cumulative effects on the return to farm labour. If the returns to farm labour were to grow at an average annual rate of about 3 percent, for example, farm prices would have to increase at least 3 percent annually (assuming other prices did not change) to have the same effect on returns to farm resources.

Costs

The costs of the agricultural price and income policies of industrial countries are substantial; they include not only direct governmental outlays but also the increased costs to consumers in those countries, as well as the losses to developing countries of potential export markets.

Organizations of Farming

Ownership

Except in the few countries with communist governments, most farmland is privately owned. That does not mean, however, that the land is owned by those who farm it. In most countries a major aspiration of farm people has been to achieve the ownership of the land they work. After World War II, for example, Japan and Taiwan underwent land reforms that were intended to broaden ownership, and similar reforms have been advocated in other countries.

On a cooperative farm the land is owned jointly by the members of the group who farm it. The cooperative generally also owns all the major means of production, and the members supply all or most of the labour. While there are examples of cooperative farms in many countries, they loom large only in Israel, where the kibbutzim control about one-tenth of all agricultural land.

In a collective farm, at least as organized in the former Soviet republics, the land was owned by the state but was permanently leased to the kolkhoz (collective farm). The kolkhoz owned its own equipment and livestock and was required to meet certain commitments to the state in the form of deliveries of farm products. In theory, the members of the kolkhoz were to elect the officers of the farm and establish the procedures by which the net product was to be divided among the members for services performed. In practice, however, their autonomy was severely limited by the economic plans. In most cases these plans were incredibly detailed, specifying the crops to be grown, the times of plowing, planting, and harvesting, the quantities of fertilizer and manures to be used, and the kinds of livestock to be maintained.

On state farms the land and all other means of production are owned by the state. The workers are paid in wages, and management decisions are made by individuals directly responsible to the state.

Kinds of Farm Operation

If a family farm is defined as one for which the farm operator and family members supply at least half of the labour, the majority of farms in the world are family farms. Family farming is carried on under a wide range of conditions, from the small farms of Asia to the highly mechanized farms of Canada, the United States, and the United Kingdom.

The family farm may be owned by the farmer or rented. The most rapidly expanding type of tenure in the United States is that in which the farmer owns part of the land and rents the remainder; almost one-third of all farmland in the United States consists of part-owner farms. This arrangement enables the farmer to increase the size of the farm through renting and to invest capital in machinery and livestock.

Family farms may be large in terms of total assets or sales. The relative importance of family farms among the largest farms in the United States has increased over the past few decades. One of the more striking changes in industrial countries has been the increased importance of nonfarm income received by farm families. In the United States, Canada, and Japan more than half of the total income of farm families comes from nonfarm sources, while in most western European countries at least a third of the income of farm families is earned outside of agriculture.

A system of tenant farming known as sharecropping developed in the South of the United States following the freeing of the slaves in the 19th century. It was essentially an adjustment of the plantation system created to permit the owners to maintain a large measure of control over farm operations. The sharecroppers usually supplied only the labour, while the owners provided animal power, machinery, and most of the other inputs in the form of an advance. The sharecroppers received what was left after they had paid back the owners—generally about half of what had been produced.

For various reasons, including the exodus of blacks from American agriculture, the introduction of farm machinery, and the reduction in the acreage of cotton, the number of sharecroppers in the South has diminished drastically since 1935.

In the second half of the 20th century, there began a growth of large-scale farming run as a business enterprise. Such "industrial farms" are of growing significance in world agriculture. There are farms covering extensive areas of land in Africa, South America, Australia, and the United States, where farms became larger as their numbers grew smaller. Such large farms tend to specialize in the production of vegetables, fruits, cotton, poultry and poultry products, and livestock.

Comparative Strengths and Weaknesses

If they were free to choose, most farm families would want to own the land they farm. Wherever collectivization of private farmers has been carried out, it has required the use of force or the threat of force. But if family farming is to be viable, it must function efficiently, which means that farmers must have access to adequate sources of credit; must be able to obtain fertilizers, machinery, and other equipment; and must be able to market their produce easily. Laws and institutions must be sufficiently flexible to permit the average size of farms to increase as economic growth occurs.

Collective farming did not fulfill the hopes of its early advocates. In the Soviet Union the collective farm was used by Joseph Stalin as a means of exploiting the rural population in order to finance the expansion of industrialization. In the post-Stalin era the incomes of collective farm members increased, and it was believed that many remaining difficulties could be eliminated if the farms were given greater freedom in running their affairs. Nothing in the concept of the collective farm required the imposition of delivery quotas, centralized control of farm investment, or a particular organization of farm labour. Another weakness of collective farms was the failure to provide adequate incentives for individual members. Because of the difficulties involved in rewarding members for their individual work on the common land, the household plots of the members all too often tended to flourish at the expense of the collective.

There is no ideal form of organization that fits all farming. Under some circumstances the ownership of land may absorb so much capital that other investments, such as machinery and livestock, are neglected. Land rental may be a better alternative for many families, especially those with limited capital. The Israeli kibbutz has made it possible for many people with little or no agricultural experience to learn farming techniques quickly and efficiently. The most important consideration is whether the other institutions economic, political, and social are adequate to provide farmers with a wide range of resources and alternatives.

Agroecology

The term agroecology can be used in multiple ways, as a science, as a movement and as a practice. Broadly stated, it is the study of the role of agriculture in the world. Agroecology provides an interdisciplinary framework with which to study the activity of agriculture. In this framework, agriculture does not exist as an isolated entity, but as part of ecology of contexts. Agroecology draws upon basic ecological principles for its conceptual framework.

Ecological Strategy

Agroecologists study a variety of agroecosystems, and the field of agroecology is not associated with any one particular method of farming, whether it be organic, conventional, intensive or extensive. Furthermore, it is not defined by certain management practices, such as the use of natural enemies in place of insecticides, or polyculture in place of monoculture.

Additionally, agroecologists do not unanimously oppose technology or inputs in agriculture but instead assess how, when, and if technology can be used in conjunction with natural, social and human assets. Agroecology proposes a context- or site-specific manner of studying agroecosystems, and as such, it recognizes that there is no universal formula or recipe for the success and maximum well-being of an agroecosystem.

Instead, agroecologists may study questions related to the four system properties of agroecosystems: productivity, stability, sustainability and equitability. As opposed to disciplines that are concerned with only one or some of these properties, agroecologists see all four properties as interconnected and integral to the success of an agroecosystem. Recognizing that these properties are found on varying spatial scales, agroecologists do not limit themselves to the study of agroecosystems at any one scale: farm, community, or global.

Agroecologists study these four properties through an interdisciplinary lens, using natural sciences to understand elements of agroecosystems such as soil properties and plant-insect interactions, as well as using social sciences to understand the effects of farming practices on rural communities, economic constraints to developing new production methods, or cultural factors determining farming practices.

Various Approaches to Agroecology

Agroecologists do not always agree about what agroecology is or should be in the long-term. Different definitions of the term agroecology can be distinguished largely by the specificity with which one defines the term "ecology," as well as the term's potential political connotations. Definitions of agroecology, therefore, may be first grouped according to the specific contexts within which they situate agriculture. Agroecology is defined by the OECD as "the study of the relation of agricultural crops and environment." This definition refers to the "ecology" part of "agroecology" narrowly as the natural environment. Following this definition, an agroecologist would study agriculture's various relationships with soil health, water quality, air quality, meso and micro-fauna, surrounding flora, environmental toxins, and other envirnomental contexts.

A more common definition of the word can be taken from Dalgaard et al., who refer to agroecology as the study of the interactions between plants, animals, humans and the environment within agricultural systems. Consequently, agroecology is inherently multidisciplinary, including factors from agronomy, ecology, sociology and economics. In this case, the "ecology" portion of "agroecology is defined broadly to include social, cultural, and economic contexts as well.

Agroecology is also defined differently according to geographic location. In the global south, the term often carries overtly political connotations. Such political definitions of the term usually ascribe to it the goals of social and economic justice; special attention, in this case, is often paid to the traditional farming knowledge of indigenous populations. North American and European uses of the term sometimes avoid the inclusion of such overtly political goals. In these cases, agroecology is seen more strictly as a scientific discipline with less specific social goals.

Fred Buttel makes a more academic distinction of the various approaches within the field, separating it into five broad categories:

1. Ecosystems Agroecology: This approach is driven by the ecosystems biology of Eugene Odum. This approach is based in the hypotheses that the natural systems, with its stability and resilience, provide the best model to mimic if sustainability is the goal. Normally, ecosystems agroecology is not actively involved in social science; however, this school is essentially based on the belief that large-scale agriculture is inappropriate. The work of Steve Gliessman is prototypical of this approach.

2. Agronomic Ecology: The basic approach in this branch is derived mostly from agronomy, including the traditional agricultural production sciences. This approach also does not actively involve social sciences in the agroecological analysis, but uses social sciences to understand the processes by which agriculture became unsustainable. Chuck Francis, Richard Hardwood, Ricardo Salvador, and Matt Liebman are exemplars of this approach.

3. Ecological Political Economy: The driving force behind this form of agroecology is a political-economical critique of modern agriculture. The school believes that only radical changes in political economy and the moral economy of research will reduce the negative costs of modern agriculture. The works of Miguel Altieri (ecosystem biologist), John Vandermeer (population ecologist), Richard Lewontin, and Richard Levins provide examples of this politically charged and socially-oriented version of agroecology.

4. Agro-Population Ecology: This approach is derived from the science of ecology primarily based on population ecology, which over the past three decades has been displacing the ecosystems biology of Odum. Buttel explains the main difference between the two categories, saying that "the application of population ecology to agroecology involves the primacy not only of analyzing agroecosystems from the perspective of the population dynamics of their constituent species, and their relationships to climate and biogeochemistry, but also there is a major emphasis placed on the role of genetics." David Andow and Alison Power are cited as examples of professionals espousing this view.

Integrated Assessment of Multifunctional Agricultural Systems

This approach focuses on the multifunctionality of the landscape, instead of focusing solely on the agricultural enterprise. Agriculture and the food system are considered parts of an institutional complex that relates to and integrates with other social institutions. Scholars adopting this highly integrated approach, mostly Europeans, do not consider any one discipline the leader of agroecology.

Holon Agroecology

First introduced in 2007 by the soil scientist William T. Bland and the environmental sociologist Michael M. Bell of the University of Wisconsin–Madison, holon agroecology draws on Koestler's notion of a "holon" which is both part and whole and develops it with ideas of narrative, intentionality, and incompleteness or unfinalizability, within an ever-changing "ecology of contexts". In contrast to systems thinking, holon agroecology stresses seeing the agricultural endeavor as an unfinished accomplishment that is constantly adjusting itself to its many contexts and their conflicts and incommensurabilities. The farm holon represents a kind of "holding together" in order to persist through change, but a holding together that is never fully unified and worked out.

Agricultural Chemistry

Agricultural chemistry is a science concerned with ways to influence chemical and biochemical processes in soil and plants, with plant mineral nutrition and with using fertilizers and other chemical means to improve fertility and increase yield. It also addresses several other ways to increase yield, such as herbicides and stimulants for growth, and serves as the scientific basis for the introduction of chemical processes into agriculture.

Agricultural chemistry is related to both chemical and biological sciences in its aims, processes, and research topics. It is also closely linked to soil science, forestry, meteorology, plant and biochemistry, agricultural microbiology, physics and chemistry. Its main subdivisions are plant nutrition, soil and fertilizer interactions, evaluation of specific types and types of fertilizers and their methods of application, soil improvement by chemical means. For example: Application of lime or gypsum and research into and usage weed control chemicals. The science of chemistry and biochemistry in its connection to agriculture.

Importance of Agricultural Chemistry

- Chemistry is considered as one of the important tools in our daily lives. Almost everything is made up of chemistry. It is important because it helps us to understand the structure, composition, and changes of matter.
- All matters are made of chemistry. Chemistry is the science behind most of the agricultural products that we mostly use, from food production to the manufacture of products like fertilizers.
- Chemistry is concerned with compounds, both organic and inorganic, and agriculture is concerned with the production of organic products using organic and inorganic inputs. Therefore, Chemistry is an integral part of farming from molecular to organ level.
- From the basics of photosynthesis to the use of agricultural products chemistry plays a role. The advances in this practice are due only to active research carried out in the field of chemistry and then to these applications that cause the land to produce more abundantly and at the same time protect it from deterioration and misuse.

Importance of Chemistry in Agriculture can be classified as follows:

- Photosynthesis: For all agricultural products this natural process provides the basic building block. The overall process is well shown by the net equation below:

  $$CO_2 + H_2O \rightarrow C_6H_{12}O_6 + O_2$$

  To life on earth no chemical process is more important than photosynthesis. Without photosynthesis, there would not only be no plants, nor could the planet sustain any kind of life. Research in this field has contributed to understanding of the mechanism and thus optimizing the conditions for maximizing this process.

- Fertilizers: Fertilizer is any organic or inorganic substance of natural or synthetic origin applied to a soil to provide one or more plant nutrients that are essential for plant growth. Fertilizers can be divided into two categories:
    - Organic Fertilizers: These fertilizers are produced from living systems and also from animal manure, compost, fish and bone meal. These organic fertilizers are broken down into the soil by microorganisms to release their nutrients for plant use. Chemical fertilizers are less complex and have high concentrations of nitrogen, phosphorus, potassium, calcium, magnesium and Sulphur in the soil.
    - Inorganic Fertilizer: Inorganic fertilizers are the substances that are synthesized through the haber – bosch process, which result in ammonia as the end product. This ammonia is used in many fertilizers with nitrogen such as anhydrous ammonium nitrate, and urea. Now fertilizers have been developed, with slow nutrient release. Effective use of fertilizers to raise crop yields has counterbalanced land loss due to urbanization and significantly helped global population growth, it has been estimated that nearly half of the world's population is currently fed through synthetic nitrogen fertilizer usage.
- Pesticides and Insecticides: A large variety of chemicals known as pesticides are used to minimize the damage of the crops by pests. Subclasses of this are, based on their purpose, herbicides, insecticides, fungicides, rodenticides, pediculicides and biocides. More safe and greener pesticides are being developed with active research in this Field. Insecticides are poisons used to kill insects because they can spread diseases of the cattle, consume stored grain and prey on crops. But not all insects are bad, and some species of insects are required to pollinate plants to ensure seed is set. Such chemicals prevent insects and other animals from destroying their crops.

Chemistry in other Areas of Agriculture

- Plastic Pipes for Improved Irrigation: Plastic was derived from chemistry, and is commonly used in cultivation. This has massively increased irrigation resulting in a better climate in which the crops grow.
- Storage and Preservation of Agriculture Products: Sulfur dioxide is used for a longer period of time to keep the grain fresh and usable. For longer shelf life food preservatives such as sodium benzoate and salicylic acid are used. The development of new generation refrigerants. Chemicals are added to encourage fruit maturing or seed germination. Food packaging has progressed due to the material produced as a result of advances in chemistry. Agricultural chemistry has increased the diversity of the human diet and increased overall food availability, both for animals and plants.

- Food Processing: Saccharin and Sweetener Development, Vitamins and Minerals. Consumers have benefited from new technologies that have increased their food's flavour, appearance, availability and nutritional value.
- Chemicals from Agriculture Waste: Advancement in chemistry has led to the development of technologies for producing a variety of agricultural waste chemicals. A good example is the processing of bagasse alcohol which is used as a feedstock for chemicals.

Chemistry of Soil

The soil is one of Earth's life-sustaining components; it serves as physical support, a source of water and nutrients, and a medium in which materials are recycled for future generations of all living things. Technically, soil is defined as the "unconsolidated material on the surface of the earth that shows evidence of life"; it is composed of inorganic and organic materials, water, and air. It is important to distinguish between a "soil" and a "soil sample." A soil is a multilayer porous medium developed in place by the weathering of the underlying rock or weathering of some other parent material (e.g., alluvium) deposited upon that rock hundreds to thousands of years before. The overall behavior of a soil is dictated by the chemical, physical, and biological composition of each layer of the profile; the latter is the basis of the soil classification system (soil taxonomy) utilized worldwide. A soil sample, on the other hand, is a small portion, removed from the soil profile, and taken to a laboratory for analysis or experimentation. Extrapolation from laboratory data to soil behavior in the field requires caution.

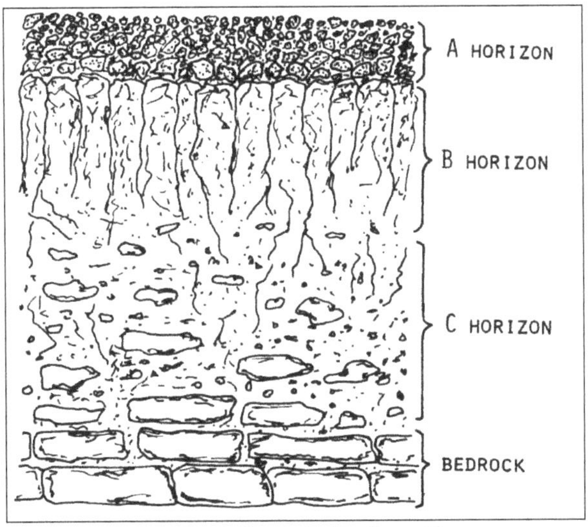

Diagram of a typical soil profile.

Soil Composition

Typical dry soil consists of 95 percent inorganic particles and about 5 percent of organic material as coatings on the inorganic particles and as separate phases. Exceptions are histosols, commonly known as peat or bog soils, which contain up to 100 percent organic matter. The inorganic material is a mixture of primary particles (sand, 2–0.02 mm diameter; silt, 0.02–0.002 mm diameter; and clay, less than 0.002 mm diameter as defined by the International Society of Soil Science), metal oxides, and salts. The elemental composition of soil is listed. Chemically, sand is quartz (SiO_2) silt consists of primary minerals such as feldspars and biotite, and clays are secondary minerals such as kaolinite, montmorillonite, and illite. The proportions of sand, silt, and clay in a soil give the soil its "texture", e.g. loam.

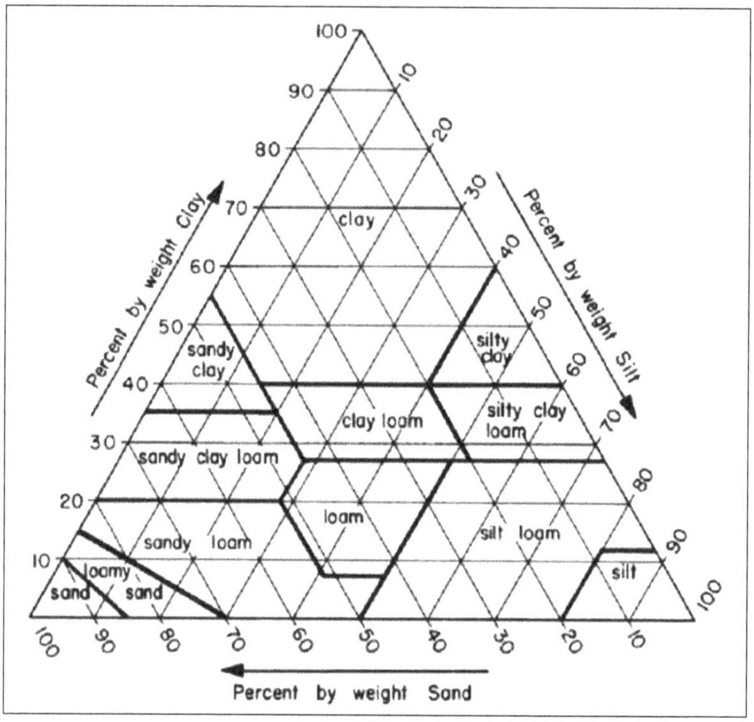

Soil textural diagram.

On the basis of particle size it can be predicted that the colloidal-size fraction (particles less than 0.002 mm diameter) will be the major site of chemical reactions in a soil due to the large specific surface (surface area per unit weight) compared to that of sand or silt fractions. Specific surface of soil particles can range from 20 to 800 $m^2 g^{-1}$ for clay, 0.04 $m^2 g^{-1}$ for silt, and 0.007 $m^2 g^{-1}$ for sand. The sand and silt fractions of a soil contribute little to chemical behavior but are important to the physical properties of a soil such as its proportion of air and water spaces (porosity), and the rate at which water can be transported from one point in the soil to another (hydraulic conductivity). The focus here is on soil chemical behavior and its basis: soil colloids and their surface reactions with the soil solution.

The Colloidal Fraction

Soil colloids include clays, metal oxides, and organic particles. The term "clay" can refer to (i) a clay soil (soil with greater than 40 percent clay particles by weight), (ii) a particle size (particles less than 0.002 mm diameter), or (iii) a specific secondary mineral (like kaolinite, montmorillonite, illite). Metal oxides and oxyhydroxides are primarily those of iron, aluminum, manganese, and titanium. Organic colloids are humic substances, that is, remains of plants and animals as well as products of microbial metabolism.

The large specific surface of soil colloids provide sites in which molecules from the soil solution can bind by Van der Waals attractive forces. Soil colloids also have electrostatic surface charge, a property that results in adsorption, binding, and exchange of ions and molecules on particle surfaces. Surface electrostatic charge originates from isomorphous substitution, ionization of functional groups, or specific adsorption, depending upon the nature of the particle.

Clays

Clays are primarily aluminosilicate secondary minerals, consisting of one to several layers of silica tetrahedra and alumina octahedra. Binding of one layer of silica tetrahedral sheet to one layer of alumina octahedral sheet yields a 1:1 clay mineral; an example is kaolinite, in which the platy minerals stack on each other tightly bonded by hydrogen bonds such that water or other ions cannot enter the layers. Such clay is termed non-expanding. The clay mineral halloysite is a 1:1 kaolin but with water molecules between each 1:1 layer. Monovalent cations can adsorb between these layers. Halloysite is tubular rather than platy.

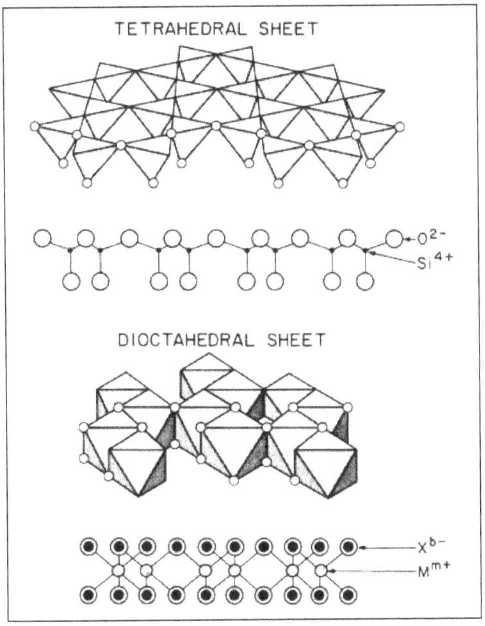

Diagram of SiO_4 tetrahedral and $Al(OH)_6$ octahedral sheets (X is hydroxyl and M is, e.g., Al). Sections with small open circles on the respective tetrahedra and octahedra are represented below each.

Clays with 2 layers of silica tetrahedral sheets with a layer of alumina octahedral sandwiched between them are termed 2:1 clay minerals, of which montmorillonite is an example. The layers of montmorillonite are held together loosely by water molecules and thus can expand to hold more water molecules and hydrated ions.

A third group are the 2:1:1 clays (the chlorites) that are 2:1 in terms of silica and alumina, but have, in addition, a hydroxide interlayer. The interlayer sheet is bound to the 2:1 clay electrostatically, and the tetrahedral layer is bound to the interlayer sheet by hydrogen bonds.

Other groups of clay minerals include the illite, mica, intergrade, interstratified, allophane, and fibrous groups. Identification of the various clay minerals in soil is usually carried out by x-ray diffraction techniques after the clay fraction has been separated from the soil and cleaned to remove surface coatings.

Surface charge on aluminosilicate clays arises typically from isomorphous substitution, that is, substitution of, for example, Al^{3+} for Si^{4+} in the tetrahedral layer, or Mg^{2+} for Al^{3+} in the octahedral layer of the clay mineral lattice without appreciable change in the shape or dimensions of the lattice. The result is a net negative charge, usually on the basal plane of the mineral; the charge is essentially constant because the lattice substitutions tend to be permanent. In some clays, such as kaolinite, there can be a variable charge at broken edges where functional groups such as O and OH can protonate or deprotonate, leaving a net positive or negative charge at that site. The hydroxide sheet of chlorites tends to have a positive charge because there are less than three OH- per Al^{3+}.

Metal Oxides and Oxyhydroxides

Table: Oxides, oxyhydroxides, and hydroxides in soils.

Aluminum oxides	Iron Oxides
Bayerite α-$Al(OH)_3$	Akageneite β-FeOOH
Boehmite γ-AlOOH	Ferrihydrite$Fe_{10}O_{15}$.$9H_2O$
Diaspore α-AlOOH	Feroxyhyte δ-FeOOH
Gibbsite γ-$Al(OH)_3$	Goethite α-FeOOH
	Hematite α-Fe_2O_3
	Lepidocrocite γ-FeOOH
	Maghemite γ -Fe_2O_3
	Magnetite Fe_3O_4
Manganese Oxides	**Titanium Oxides**
Birnessite δ-MnO2	Anatase TiO2
Pyrolusite β-MnO_2	Rutile TiO_2

Oxides of aluminum, iron, and manganese do not typically occur in large quantities in temperate region soils but can be present to a greater extent in tropical and highly weathered soils. They have high specific surface and react readily with ions in the soil solution. They can occur as discrete solid phases, as coatings on clays and organic colloids, and as mixed gels. The most common aluminum oxides are gibbsite and boehmite, an oxyhydroxide. The most common iron oxide is goethite. Birnessite, the most common oxide of manganese in soil, acts as a natural oxidant for certain metals, e.g. As^{3+} and Cr^{3+}.

Surface charge on metal oxides usually results from ionization of functional groups, mainly hydroxides and oxyhydroxides, resulting in surfaces with variable charge whose sign is reversible and depends on protonization and deprotonization of the sites. The pH of the surrounding medium and the pH of zero charge (PZC) of the particle affect the sign and magnitude of the charge. Another source of surface charge typical of metal oxides and oxyhydroxides is termed specific adsorption. For example, an anion such as orthophosphate, HPO_4^{2-}, can bind as an inner-sphere complex (no intervening water molecules between the ion and the surface) to a surface hydroxyl group of an oxide particle (iron oxide, for example), resulting in negative charge at that site.

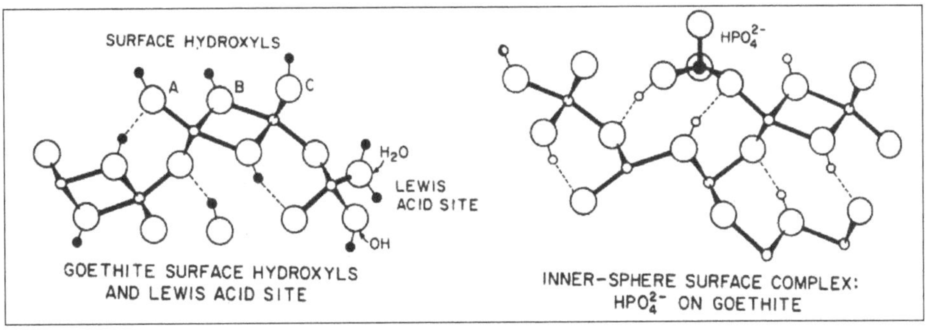

Specific adsorption of orthophosphate on iron oxide.

Organic Colloids

Soil organic matter can occur as soluble organic compounds or as organic particulate phases, commonly known as humus, and/or as coatings on clays and metal oxides. Humus includes humic substances, non-humic substances, and products of microbial metabolism. Soil organic colloids have high specific surface (up to 900 m^2g^{-1}) and large amounts of negative surface charge (150 to 300 cmol kg^{-1}) due to ionization of functional groups such as carboxyl, phenolic, alcoholic, and carbonyl. This material thus can bind, or is a source of, plant nutrients, metal ions, and xenobiotic compounds (e.g. pesticides and other contaminants). Due to the variable nature of organic matter in soil, identification of individual compounds is not routinely done; rather, a definition of soil organic matter groups is determined operationally.

The Soil Solution

The soil solution refers to the aqueous phase permeating the pores of a soil; its chemical behavior is governed by the rules of aqueous chemistry except that the presence of colloids, both organic and inorganic, adds reactions involving adsorption and desorption, ion exchange, mineralogical dissolution and precipitation, and ion uptake and exudation by plant roots. Reactions typical of the soil solution itself include oxidation and reduction, complex formation, dissolution and precipitation, and acid-base. Soil solutions include dissolved salts (approximately 0.02 N in a typical soil), soluble organic compounds, metal organic and inorganic complexes, enzymes and other soluble biochemicals. The pH of typical soil solutions range from 4 to 9, and govern the mobility and availability of nutrients and contaminants to plants as well as enzymatic activity of soil microorganisms.

Soil Solution Effects on Particle Surface Charge

The surface charge on a colloid results in attracting or repelling ions in the adjacent solution phase, and thus governs the nature and extent of ionic and molecular binding to the particle. Regardless of the nature of the particle, it can be modeled as a solid sphere, with a uniform charge distributed over its surface, surrounded by water containing ions in solution. Distribution of these ions on and around the particle can be quantified by "double layer" theory, a simplified version of which is illustrated here.

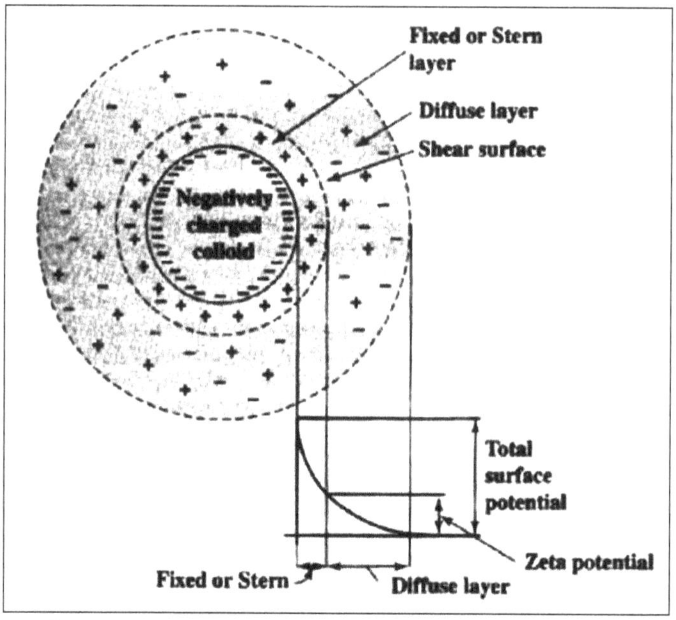

Schematic diagram of the double layer and electric potential around a negatively charged colloidal particle.

Ions with charge opposite to that of the surface are strongly bound to the surface by the Coulombic force, and are called the "fixed" or Stern layer. Outside the Stern layer, the ions are more diffuse (the Guoy-Chapman layer). As the Coulombic force decreases with the square of the distance from the particle surface, ions of opposite charge under Brownian motion in the solution, are more diffusely attracted, while ions of the same charge as the particle surface are repelled with decreasing force. At infinite distance (the bulk solution), the cations and anions are at electroneutrality. Although this is not the whole story (there are several models of double-layer theory), this model may help to understand some aspects of adsorption and ion exchange in soil. One difference between this simplified model and more refined models, for example, is the identification of "inner sphere" and "outer sphere" complex charges. The inner sphere complex charge arises from inner-sphere complexes, i.e. ion-pairs that have direct, short-range bonding between the ion and the particle, with no water molecules in between. Outer sphere complex charge arises from outer-sphere complexes, i.e. ion pairs that have at least one water of hydration between the ion and the particle; bonding is Coulombic and long-range.

The total surface electric potential due to the particle charge cannot be measured directly but can be calculated if surface charge density (electrostatic charge per unit area of particle) is known. The zeta potential, which is sometimes used as an approximation of the surface potential, can be experimentally determined by measuring the electrophoretic mobility of the particle in an electric field. The extent (thickness) of the double layer (x in cm) can be estimated from the ionic strength (I) of the solution, i.e.

$$x \approx 2.8 \times 10^{-8} \times 1^{-0.5}$$

Thus, in a typical soil solution, with ionic strength about 0.02 M, the thickness of the double layer is estimated to be 2 nm. In high-ionic strength solutions such as sea water (~ 0.7 M), the diffuse layer is about 0. Nm thick.

Solution pH and Points of Zero Charge

Table: Approximate values of points of zero charge (PZC) of soil minerals. Values are not necessarily comparable, as they were determined by different investigators under different conditions.

Mineral	PZC	Mineral	PZC
α-Al_2O_3	9.1	β-MnO_2	7.2
α-$Al(OH)_3$	5.0	SiO_2	2.0
γ-AlOOH	8.2	$ZrSiO_4$	5.0
CuO	9.5	Feldspars	2-2.4
Fe_3O_4	6.5	Kaolinite	4.6
$Fe(OH)_3$ amorphous	8.5	Montmorillonite	2.5
MgO	12.4	Albite	2.0
δ-MnO_2	2.8	Chrysotile	>10

The sign and magnitude of the surface charge on a soil colloid is determined by the nature of the particle and by the pH of the solution surrounding the particle particularly in the case of metal oxides. Understanding this phenomenon is essential to predict the fate and transport of contaminants in soil, e.g. if it is assumed that Pb^{2+} in the soil will be bound tightly to the soil due to the assumed negative charge on the soil particles, the prediction will be incorrect if there is charge reversal on clays and metal oxides in the soil under the pH conditions of the soil solution. Approximate points (pH) of zero charge of common soil minerals are listed in table.

If the pH of the colloid suspension is equal to the PZC (pH of zero charge), then the particles have no net charge; if pH < PZC then the particles have a net positive charge; if pH > PZC, then the particles have a net negative charge.

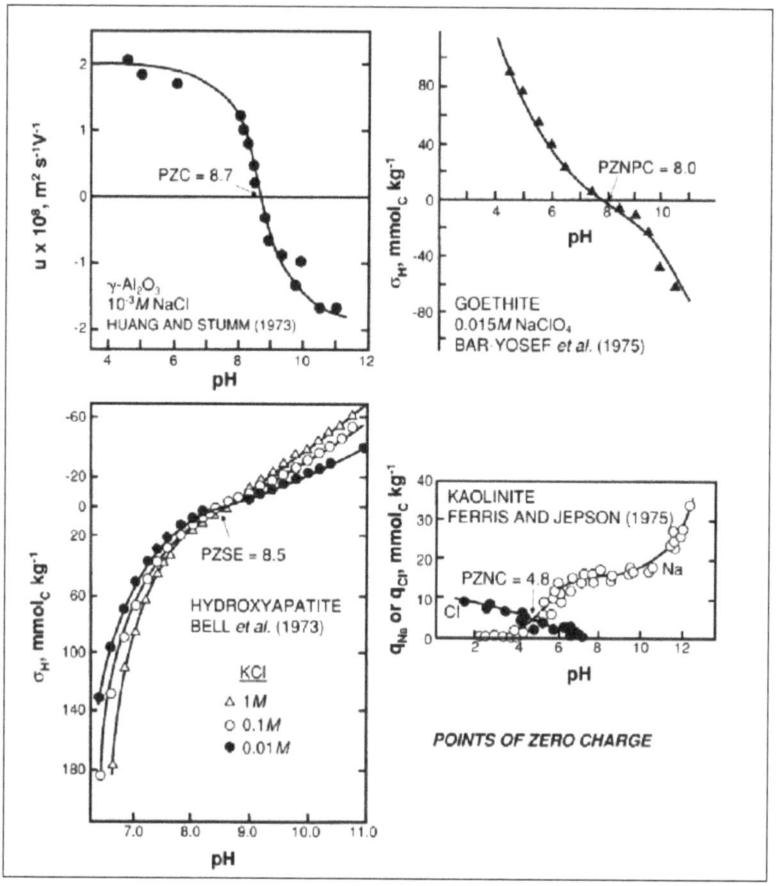

Examples of various points of zero charge. μ (top left) is electrophoretic mobility.

There is often some confusion in regard to nomenclature on zero points of charge, and can be clarified by defining four components that make up the total net charge on a particle, σ_p.

$$\sigma_p = \sigma_o + \sigma_H + \sigma_{IS} + \sigma_{OS}$$

Where, σ_o is permanent charge due to structural substitutions, σ_H is net proton charge (difference between the moles of protons and moles of OH ions complexed by surface functional groups, σ_{IS} is inner-sphere complex charge, and σ_{OS} is outer-sphere complex charge. The total net charge on a particle must be balanced by the charge of the diffuse swarm of ions in the surrounding solution, σ_D, such that the sum of the two is zero. Important pH values can now be defined.

- PZC is point of zero charge equal to the pH at which $\sigma_p = 0$.
- PZNPC is point of zero net proton charge equal to the pH at which $\sigma_H = 0$.
- PZNC is point of zero net charge equal to the pH at which $\sigma_{IS} = \sigma_{OS} + \sigma_D = 0$, i.e. the pH at which the difference between the cation exchange capacity and anion exchange capacity is zero.
- PZSE is point of zero salt effect equal to the pH at which the potentiometric titration curves intersect regardless of the background solution ionic strength.

Ion Exchange in Soil

Soil colloids are the seat of reaction. Ion exchange (cation and anions) takes place in colloids. The phenomenon of ion exchange is of great importance in agriculture. It has considerable influence on the liberation of plant nutrients such as Ca, K, P, etc. It controls soil structure and crumb formation. It is also responsible for imparting a stable structure. It controls the processes plays an important role in the reclamation of acid and alkali soils. It also influences the effect of fertilizers and fertilizer practices.

Ion exchange is two types: (1) cation exchange or base exchange and (2) anion exchange or acid exchange. Ion exchange is a reversible process in which cation and anion exchanged between solid and liquid phase.

Cation Exchange or Base Exchange

In a near neutral soil, Ca is remaining adsorbed on colloidal particle H ion generated as organic and mineral acid formed due to decomposition of organic matter. In colloid, H ion is adsorbed more strongly than is the Ca and H is chemically equivalent. The reaction is as follow:

$$\text{Ca}-\text{Clay micelle}+2\text{H}+ \rightleftarrows \begin{matrix}\text{H}\\ \text{H}\end{matrix}>\text{Clay micelle}+\text{Ca}^{2+}$$

This phenomenon of the exchange of cations between soil and salt solution is known as calion exchange or Base Exchange and the cations that take part in this reaction are called exchangeable cations. It is the interchange between a cation in solution and another cation on the surface of clay or organic matter (colloids). Thompson and Way first recognised cation exchange phenomenon. They passed a solution of $(NH_4)_2SO_4$

through a soil column and found that effluent contained no $NH4^-$ salt but $CaSO_4$. This phenomenon is always in equivalent quantities of ions.

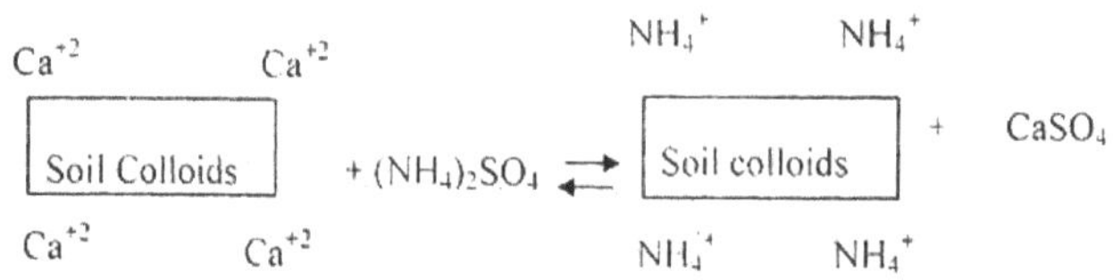

Types of Silicate Clay Mineral

In case of kaolinite, monovalent cations are more strongly held than divalent, while in case of montmorillonite, divalent cations are strongly held than monovalent cations. The normal order of adsorbed cations on clay particles is:

$$H^+ > Al^{+3} > Ca^{+2} = Mg^{+2} > K^+ = NH_4^+ > Na^+.$$

Significance of Cation Exchange

Cation exchange is an important reation in soil fertility, in causing and correcting soil acidity and basicity. In changes altering soil physical properties, and as a mechanism in purifying or altering percolating Waters. The plant nutrients like calcium, magnesium and potassium are supplied to plants in large measure from exchangeable forms. Cation exchange is very important in soils because of the following relationships:

- The exchangeable K is a major source of plant K.
- The exchangeable Mg is often, a major source of plant Mg.
- The amount of lime required to raise the pH of an acidic soil is greater as the CEC is greater.
- Cation exchange sites hold Ca^{+2}, Mg^{+2}, K^+, Na^+ and NH_4^+ ions and slow down their losses by leaching.
- Cation exchange sites hold- fertilizer K^+ and NH_4^+ and greatly reduces their mobility in soils.
- Cation exchange sites adsorb many metals (Cd^{+2}, Zn^{+2}, Ni^{+2} and Pb^{+2}) that might be present in wastewater. Adsorption removes them from the percolating water, thereby cleansing the water that drains into groundwater.

Anion Exchange

The process of anion exchange is similar to that of cation exchange. Under certain conditions hydrous oxides of iron and aluminium shows evidence of having positive charges on their crystal surfaces. The positive charges of colloids are due to addition of hydrogen (H^+) in hydroxyl group (OH^-) resulted in net positive charge (OH^{2+}). This positive charge will attract anions. The capacity for holding anions increases with increase in acidity. The lower the pH the greater is the adsorption. All anions are not adsorbed equally readily. Some anions such as H_2PO_4 are adsorbed very quickly at all pH values in the acid as well as alkaline range, Cl^- and SO_4^{--} ions are adsorbed slightly at low pH but none at neutral soil, while NO_3^- ions are not adsorbed at all. Hence, at the pH commonly prevailing in cultivated soils- nitrate (NO_3), chloride (Cl) and sulphate (SO_4) ions are easily lost by leaching. In general, the relative order of anion exchange is:

$$OH^- > H_2PO_4^- > SO_4^- > NO_3^-$$

$$R.OH + HCI \rightarrow R.OH_2 + CI^-$$

$$R.NH_2 + HCI \rightarrow R.NH_3 + CI^-$$

This adsorbed negatively changed anion is replaceable as under:

$$\underset{(R=AI.Fe)}{R.OH^+_2} + \underset{\text{soil solution}}{CI^- + H_2PO_4^-} \rightarrow \underset{\text{on exchange complex}}{R.OH_2.H_2PO_4} + \underset{\text{soil solution}}{CI^-}$$

In this way there is fixation of phosphate ion, in which OH ion of silicate clay mineral is substitute by H_2PO_4 ion by isomorphrous substitution. The H_2PO_4 ion becomes a part of the silicate clay minerals and the phosphate fixed by this way is known as colloid bound phosphate. This phosphate is not exchangeable and available to the plant. The anions like NO_3 and OH are not bound at pH above 7. The adsorbing power of anions is in the order of:

$$CI^* = NO_3^* < SO_4 < PO_4$$

* are not held by the soil colloids and are lost due to the leaching.

Or $OH > H_2PO_4 > SO_4 > NO_3$ (relative order of anion exchange).

Importance of Anion Exchange in Nutrient Availability

The phenomenon of anion exchange assumes in relation to phosphate ions and their fixation. The exchange is brought mainly by the replacement of OH ions of the clay minerals.

$$OH\,Clay + \underset{\text{Available}}{H_2PO_4} \rightarrow \underset{\text{unavailable}}{H_2PO_4\,Clay} + OH$$

The adsorption of phosphate ions by clay particles from soil solution reduces its availability to plants. This is also known as colloid bound phosphate fixation. The phosphate ion again becomes available when lime is applied to increase the pH of acidic soil.

The OH ion originated not only from silicate clay minerals but also from hydrous oxides of iron and aluminum present in the soil. The phosphates ions react with the hydrous oxides also get fixed forming insoluble hydroxyl phosphate of ion and aluminum. Which is called saloid bound phosphate.

$$AI(OH)_3 + \underset{\text{Available}}{H_2PO_4^-} \rightarrow \underset{\text{unavailable}}{AI(OH)_2\, H_2PO_4} + OH^-$$

If the reaction takes place at a low pH under strongly acid conditions, the phosphate ions are irreversibly fixed and are totally unavailable for the use of plants.

Agrochemicals

Agrochemicals or agrichemicals, is a common name given to chemicals which are used in agriculture, to aid plants and crops growth and safety. Agrochemicals were manufactured to protect agricultural crops from pests and for augmenting crop yields. It is a costly input for agriculture.

Classifications of Agrochemicals

In most of the cases, agrochemicals refer to pesticides which include insecticides, herbicides, fungicides, rodenticides, molluscicides, and nematicides. Agrochemical also includes fertilizers and soil conditioners.

- Pesticides: Pesticide is a chemical or a substance used to destroy or control some types of plants or organisms also known as pests, which are harmful to cultivated plants or to animals. Pesticides mostly work through poisoning pests.
- Insecticides: It is used to destroy insects. Insecticides can be ovicides that kill eggs, larvicides to kill larvae. Pesticides examples: Organochlorines, organophosphates, carbamates, and pyrethroids.
- Herbicides: It is used to control or kill weeds and herbs. Herbicides examples: Gramoxone and glyphosate.
- Fungicides: It is used for controlling fungi and oomycetes. Fungicides examples: Mankocide.
- Algaecides: Used for controlling algae. Also known as algicides.
- Rodenticides: It is used to prevent spread of rodents like rat, mice. Examples: Klerat.
- Molluscicides: Used to control molluscs like snails and slugs. Examples: Slugit.
- Nematicides: It is used to control or killing nematodes. Examples: Furadan.
- Fertilisers: These are chemical compounds used for promoting plant growth. They are used to mitigate nutrient deficiency in the soil. Typically, it is

applied to soils or to plant tissues. Fertilisers can be categorized into two categories: organic and inorganic fertilisers. Organic fertilisers are naturally existing substances prepared through natural processes. Inorganic fertilisers, also referred to as synthetic fertilizers are manufactured artificially using chemical processes by utilizing natural deposits, which are altered chemically.

- Soil conditioners: To keep all soils in good condition, the best thing to do is to add things that keep it in good condition. These good things are called a soil conditioner that includes manure, compost, peat, livestock manures and leaves. All these things are laid on top of the soil and then mixed. Soil conditioners enhance aeration and water holding capacity of the soil.
- Liming and acidifying agents: Soils sometimes can be too acidic or too alkaline for proper growth of crops. In these cases, liming and acidifying products are added to soil to adjust its pH. When the soil is too acidic, calcite on the form of powdered limestone is added primarily, whereas for more alkaline soil sulfur compounds are added to neutralize.

Agrochemicals and their Effects

Although agrochemicals are initially used to improve crop's health, overuse of these chemicals has now started affecting the environment in many ways. Excessive utilization of these chemicals generates residues that cause nutrients imbalance and reduction of the yielding capacity of crops.

Effects on Soil:

- May kill beneficial bacteria.
- Increase nitrate level in soils.
- Alter the pH of the soil.
- Residual effect.
- Kill soil organisms.
- Toxicity and reduction in soil quality.

Effects on Water:

- Water becomes unfit for consumption.
- Can promote the growth of algae.
- Leads to eutrophication due to excessive chemicals.
- Affect aquatic animals due to water pollution.

Effects on Air:

- Residues and particles of these chemicals can lead to air pollution.
- Pesticides drift happens when air carries agrochemicals particles from one place to other.
- More amount of spray evaporation may happen due to low relative humidity and higher temperatures.
- Effect surrounding organisms' health due to inhalation of polluted air.

Effects on Human Health:

- It causes variety of health effects, from simple skin and eyes irritation.
- It also affects the nervous system, causes cancer and also reproductive problems.
- Can cause nerve damage, infertility, hormones disorders and neurotoxicity.

Manure spreader.

Manure is the organic material that is used to fertilize land, usually consisting of the feces and urine of domestic livestock, with or without accompanying litter such as straw, hay, or bedding. Farm animals void most of the nitrogen, phosphorus, and potassium that is present in the food they eat, and this constitutes an enormous fertility resource. In some countries, human excrement is also used. Livestock manure is less rich in nitrogen, phosphorus, and potash than synthetic fertilizers and hence must be applied in much greater quantities than the latter. A ton of manure from cattle, hogs, or horses usually contains only 10 pounds of nitrogen, 5 pounds of phosphorus pentoxide, and 10 pounds of potash. But manure is rich in organic matter, or humus, and thus improves the soil's capacity to absorb and store water, thus preventing erosion. Much of the potassium and nitrogen in manure can be lost through leaching if the material is exposed to rainfall before being applied to the field. These nutrient losses may be prevented by such methods as stacking manure under cover or in pits to prevent leaching, spreading it on fields as soon as it is feasible, and spreading preservative materials in the stable. A green manure is a cover crop of some kind, such as rye, that is plowed under while still green to add fertility and conditioning to the soil.

The use of manure as fertilizer dates to the beginnings of agriculture. On modern farms manure is usually applied with a manure spreader, a four-wheeled self-propelled or two-wheeled tractor-drawn wagon. As the spreader moves, a drag-chain conveyor located at the bottom of the box sweeps the manure to the rear, where it is successively shredded by a pair of beaters before being spread by rotating spiral fins. Home gardeners like to use well-rotted manure, since it is less odorous, more easily spread, and less likely to "burn" plants.

Manures and Fertilizers

Fertilizer is a natural or artificial substance containing the chemical elements that improve growth and productiveness of plants. Fertilizers enhance the natural fertility of the soil or replace the chemical elements taken from the soil by previous crops.

The use of manure and composts as fertilizers is probably almost as old as agriculture. Modern chemical fertilizers include one or more of the three elements that are most important in plant nutrition: nitrogen, phosphorus, and potassium. Of secondary importance are the elements sulfur, magnesium, and calcium.

Most nitrogen fertilizers are obtained from synthetic ammonia; this chemical compound (NH_3) is used either as a gas or in a water solution, or it is converted into salts such as ammonium sulfate, ammonium nitrate, and ammonium phosphate, but packinghouse wastes, treated garbage, sewage, and manure are also common sources of it. Phosphorus fertilizers include calcium phosphate derived from phosphate rock or bones. The more soluble superphosphate and triple superphosphate preparations are obtained by the treatment of calcium phosphate with sulfuric and phosphoric acid, respectively. Potassium fertilizers, namely potassium chloride and potassium sulfate, are mined from potash deposits. Mixed fertilizers contain more than one of the three major nutrients—nitrogen, phosphorus, and potassium. Mixed fertilizers can be formulated in hundreds of ways.

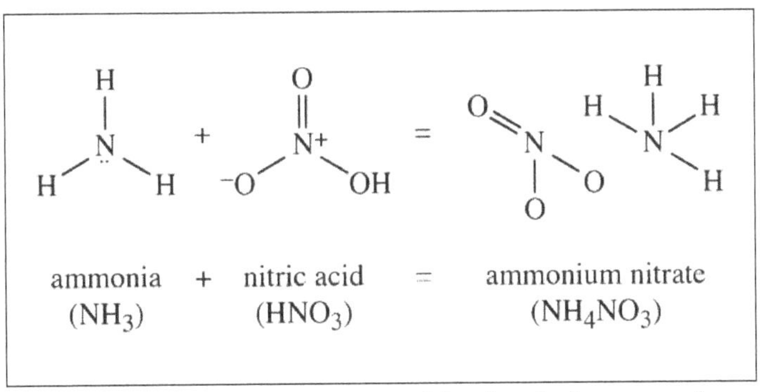

Ammonium nitrate: Structure of the chemical compound ammonium nitrate, from which fertilizer is produced.

On modern farms a variety of machines are used to apply synthetic fertilizer in solid, gaseous, or liquid form. One type distributes anhydrous ammonia, a liquid under pressure, which becomes a nitrogenous gas when freed from pressure as it enters the soil. A metering device operates valves to release the liquid from the tank. Solid-fertilizer distributors have a wide hopper, with holes in the bottom; distribution is effected by various means, such as rollers, agitators, or endless chains traversing the hopper bottom. Broadcast distributors have a tub-shaped hopper from which the material falls onto revolving disks that distribute it in a broad swath.

References

- Agricultural-sciences, science: britannica.com, Retrieved 11, July 2020
- What-is-an-agricultural-product-2538211: thebalancesmb.com, Retrieved 26, March 2020
- Agriculture-agricultural-practices, biology: byjus.com, Retrieved 09, June 2020
- Understanding-agricultural-systems, chapter-one-introduction-to-agricultural-systems-and-agricultural-systems-research-a-paradigm-change, systems-research-for-agriculture: sare.org, Retrieved 13, January 2020
- Soil-the-foundation-of-agriculture, knowledge, scitable-84224268: nature.com, Retrieved 21, May 2020
- What-is-agroecology, what-is-ecology-296: environment-ecology.com, Retrieved 19, February 2020
- Agricultural-chemistry: vedantu.com, Retrieved 25, August 2020
- Agrochemicals-types-and-their-effects-629: worldofchemicals.com, Retrieved 02, April 2020
- Fertilizer, topic: britannica.com, Retrieved 05, January 2020

Chapter 2

Agricultural Engineering

The field of study that is concerned with the application of engineering science for designing principles for agricultural purposes is known as agricultural engineering. It is a multidisciplinary field that combines various disciplines of environmental, mechanical, food science and chemical engineering to improve the efficiency of farms. All these diverse facets of agricultural engineering have been carefully analyzed in this chapter.

"Agricultural Engineering" is one of the disciplines of the engineering. This discipline of engineering is emerging faster than the other programmes. Agricultural engineering consists of two words as 'Agricultural' and 'engineering'.

'Agricultural' stands for harvesting and husbandry. It is the cultivation of plants for foods, medicine and bio fuel and other substance that is useful for the human life.

'Engineering' is the application that involves practical and scientific knowledge in term of the machine, process, invent, design, maintain, material, system, device, built and improve.

The use of mechanized power and machinery on the farm has increased greatly throughout the world, fourfold in the United States since 1930. Research in energy use, fluid power, machinery development, laser and microprocessor control for maintaining grain quality, and farm structures is expected to result in further gains in the efficiency with which food and fibre are produced and processed.

Agricultural production presents many engineering problems and opportunities. Agricultural operations—soil conservation and preparation; crop cultivation and harvesting; animal production; and commodities transportation, processing, packaging, and storage—are precision operations involving large tonnages, heavy power, and critical factors of time and place. Facilities designed to aid farm operations help farm workers to minimize the time and energy requirements of routine jobs.

Four primary branches have developed within agricultural engineering, based on the problems encountered. Farm power and machinery engineering is concerned with advances in farm mechanization—tractors, field machinery, and other mechanical equipment. Farm structures engineering studies the problems of providing shelter for animals and human beings, crop storage, and other special-purpose facilities. Soil and water control engineering deals with soil drainage, irrigation, conservation, hydrology, and flood control. Electric power and processing engineering is concerned with the distribution of electric power on the farm and its application to a variety of uses, such as lighting to control plant growth and certain animal production operations.

Agricultural Technology

Agricultural technology is the application of techniques to control the growth and harvesting of animal and vegetable products.

Soil Preparation

Mechanical processing of soil so that it is in the proper physical condition for planting is usually referred to as tilling; adding nutrients and trace elements is called fertilizing. Both processes are important in agricultural operations.

Tilling

Tillage is the manipulation of the soil into a desired condition by mechanical means; tools are employed to achieve some desired effect (such as pulverization, cutting, or movement). Soil is tilled to change its structure, to kill weeds, and to manage crop residues. Soil-structure modification is often necessary to facilitate the intake, storage, and transmission of water and to provide a good environment for seeds and roots. Elimination of weeds is important, because they compete for water, nutrients, and light. Crop residues on the surface must be managed in order to provide conditions suitable for seeding and cultivating a crop.

Generally speaking, if the size of the soil aggregates or particles is satisfactory, preparation of the seedbed will consist only of removing weeds and the management of residues. Unfortunately, the practices associated with planting, cultivating, and harvesting usually cause destruction of soil structure. This leaves preparation of the seedbed as the best opportunity to create desirable structure, in which large and stable pores extend from the soil surface to the water table or drains, ensuring rapid infiltration and drainage of excess or free water and promoting aeration of the subsoil. When these large pores are interspersed with small ones, the soil will retain and store moisture also.

Seedbed-preparation procedures depend on soil texture and the desired change in size of aggregates. In soils of coarse texture, tillage will increase aggregate size, provided it is done when only the small pores are just filled with water; tillage at other than this ideal moisture will make for smaller aggregates. By contrast, fine-textured soils form clods; these require breakage into smaller units by weathering or by machines. If too wet or too dry, the power requirements for shattering dry clods or cutting wet ones are prohibitive when using tillage alone. Thus, the farmer usually attempts tillage of such soils only after a slow rain has moistened the clods and made them friable.

Some soils require deepening of the root zone to permit increased rate of water intake and improved storage. Unfavourable aeration in zones of poor drainage also limits root development and inhibits use of water in the subsoil.

Tillage, particularly conventional plowing, may create a hardpan, or plow sole; that is, a compacted layer just below the zone disturbed by tillage. Such layers are more prevalent with increasing levels of mechanization; they reduce crop yields and must be shattered, allowing water to be stored in and below the shattered zone for later crops.

Primary Tillage Equipment

Equipment used to break and loosen soil for a depth of six to 36 inches (15 to 90 centimetres) may be called primary tillage equipment. It includes moldboard, disk, rotary, chisel, and subsoil plows.

The moldboard plow is adapted to the breaking of many soil types. It is well suited for turning under and covering crop residues. There are hundreds of different designs, each intended to function best in performing certain tasks in specified soils. The part that breaks the soil is called the bottom or base; it is composed of the share, the landside, and the moldboard.

When a bottom turns the soil, it cuts a trench, or furrow, throwing to one side a ribbon of soil that is called the furrow slice. When plowing is started in the middle of a strip of land, a furrow is plowed across the field; on the return trip, a furrow slice is lapped over the first slice. This leaves a slightly higher ridge than the second, third, and other slices. The ridge is called a back furrow. When two strips of land are finished, the last furrows cut leave a trench about twice the width of one bottom, called a dead furrow. When land is broken by continuous lapping of furrows, it is called flat broken. If land is broken in alternate back furrows and dead furrows, it is said to be bedded or listed.

Different soils require different-shaped moldboards in order to give the same degree of pulverization of the soil. Thus, moldboards are divided into several different classes, including stubble, general-purpose, general-purpose for clay and stiff-sod soil, slat, blackland, and chilled general-purpose. The blackland bottom is used, for example, in areas in which the soil does not scour easily; that is, where the soil does not leave the surface of the emerging plow clean and polished.

The share is the cutting edge of the moldboard plow. Its configuration is related to soil type, particularly in the down suction, or concavity, of its lower surface. Generally, three degrees of down suction are recognized: regular for light soil, deep for ordinary dry soil, and double-deep for clay and gravelly soils. In addition, the share has horizontal suction, which is the amount its point is bent out of line with the landside. Down suction causes the plow to penetrate to proper depth when pulled forward, while horizontal suction causes the plow to create the desired width of furrow.

Moldboard-plow bottom sizes refer to width between share wing and the landside. Tractor-plow sizes generally range from 10 to 18 inches (25 to 45 centimetres), although larger, special-purpose types exist.

On modern mechanized farms, plow bottoms are connected to tractors either as trailing implements or integrally. One or more bottoms may be so attached. They are found paired right and left occasionally (two-way), with the advantage of throwing the furrow slice in a constant direction as the turns are made. A variation is the middlebreaker, or lister, which is a bottom equipped with both right- and left-handed moldboards.

The disk plow employs round, concave disks of hardened steel, sharpened and sometimes serrated on the edge, with diameters ranging from 20 to 38 inches (50 to 95 centimetres). It reduces friction by making a rolling bottom in place of a sliding one. Its draft is about the same as that of the moldboard plow. The disk plow works to advantage in situations where the moldboard will not, as in sticky nonscouring soils; in fields with a plow sole; in dry, hard ground; in peat soils; and for deep plowing. The disk-plow bottom is usually equipped with a scraper that aids in pulverizing the furrow slice. Disk plows are either trailed or mounted integrally on a tractor.

The rotary plow's essential feature is a set of knives or tines rotated on a shaft by a power source. The knives chop the soil up and throw it against a hood that covers the knife set. These machines can create good seedbeds, but their high cost and extra power requirement have limited general adoption, except for the small garden tractor.

The chisel plow is equipped with narrow, double-ended shovels, or chisel points, mounted on long shanks. These points rip through the soil and stir it but do not invert and pulverize as well as the moldboard and disk plows. The chisel plow is often used to loosen hard, dry soils prior to using regular plows; it is also useful for shattering plow sole.

Subsoil plows are similar in principle but are much larger, since they are used to penetrate soil to depths of 20 to 36 inches (50 to 90 centimetres). Tractors of 60 to 85 horsepower are required to pull a single subsoil point through a hard soil at a depth of 36 inches. These plows are sometimes equipped with a torpedo-shaped attachment for making subsurface drainage channels.

Secondary Tillage

Secondary tillage, to improve the seedbed by increased soil pulverization, to conserve moisture through destruction of weeds, and to cut up crop residues, is accomplished by use of various types of harrows, rollers, or pulverizers, and tools for mulching and fallowing. Used for stirring the soil at comparatively shallow depths, secondary-tillage equipment is generally employed after the deeper primary-tillage operations; some primary tillage tools, however, are usable for secondary tillage. There are five principal types of harrows: the disk, the spike-tooth, the spring-tooth, the rotary cross-harrow, and the soil surgeon. Rollers, or pulverizers, with V-shaped wheels make a firm and continuous seedbed while crushing clods. These tools often are combined with each other.

When moisture is scarce and controls of wind and water erosion necessary, tillage is sometimes carried out in such a way that crop residues are left on the surface. This system is called trash farming, stubble mulch, or subsurface tillage. Principal equipment for subsurface tillage consists of sweeps and rod weeders. Sweeps are V-shaped knives drawn below the surface with cutting planes horizontal. A mounted set of sweeps provided with power lift and depth regulation is often called a field cultivator.

The typical rod weeder consists of a frame with several plowlike beams, each having a bearing at its point. Rods are extended through the bearings, which revolve slowly under power from a drive wheel. The revolving rod runs a few inches below the surface and pulls up vegetative growth; clearance of the growth from the rod is assisted by its rotation. Rod weeders are sometimes attached to chisel plows.

Some control of weeds is obtained by tillage that leaves the middles between crop rows loose and cloddy. When a good seedbed is prepared only in the row, the seeded crop can become established ahead of the weeds. Plowing with the moldboard plow buries the weed seeds, retards their sprouting, and tends to reduce the operations needed to control them. If weed infestations become bad, they can be reduced somewhat by undercutting.

Since rainfall amount and distribution seldom match crop needs, farmers usually prefer tillage methods that encourage soil-moisture storage at times when crops are not growing. From the soil-moisture standpoint, any tillage practice that does not control weeds and result in greater moisture intake and retention during the storage period is probably unnecessary or undesirable.

Minimum Tillage

The use of cropping systems with minimal tillage is usually desirable, because intensive tillage tends to break down soil structure. Techniques such as mulching also help prevent raindrops from injuring the surface structure. Excessive tillage leaves the soil susceptible to crusting, impedes water intake, increases runoff, and thus reduces water storage for crop use. Intensive vegetable production in warm climates where three crops per year may be grown on the same land may reduce the soil to a single-grain structure that facilitates surface cementation and poor aeration.

The loosening and granulating actions of plowing may improve soil structure if the plowing is done when the moisture content is optimum; if not so timed, however, plowing can create unfavourable structure. The lifting and inversion of the furrow slice likewise may not always be desirable, because in many cases it is better to leave a trashy surface.

The concept of minimum tillage has received much attention. One type of minimum tillage consists in seeding small grain in sod that has been relatively undisturbed. Narrow slits are cut in the sod and seed and fertilizer placed in the breaks thus formed. Soil normally subject to erosion can be planted to grain this way while still retaining the erosion resistance of the sod. The technique has been successful in preparing winter grazing in south-eastern portions of the United States. In another type of minimum tillage, the land is broken and planted without further tillage in seedbed preparation. One approach involves breaking the land and planting seeds in the tractor tracks (wheel-track planting); the tractor weight crushes clods and leaves the seed surrounded by firm soil. Another method consists of mounting a planter behind the plow, thus planting without further traffic and leaving a loose seedbed that is satisfactory in areas where post-planting rains may be heavy. In some areas, where winter rain often comes after wheat is drilled, a rotation of wheat following peas has been successful. After the peas have been harvested, the field is rough plowed, and fall wheat is then drilled in directly. All these methods minimize expense and land preparation, tending to leave the soil rough, which reduces erosion and increases water intake. Somewhat similar systems are employed with row crops, where chemical weed control assists in reducing need for cultivation.

Mulch Tillage

In this system, crop residues are left on the surface, and subsurface tillage leaves them relatively undisturbed. In dryland areas, a maximum amount of mulch is left on the surface; in more humid regions, however, some of the mulch is buried. Planting is accomplished with disk openers that go through several inches of mulch. Since mulch decomposition may deprive the crop of nitrogen, extra fertilizer is often placed below the mulch in humid areas. In rainy sections, intercropping extends the protection against erosion provided by mulches. Intercrops are typically small grains or sod crops such as alfalfa or clover grown between the rows of a field crop that reach maturity shortly after the field crop has been established and furnish mulch cover for a long time.

If growth of the intercrop competes with the main crop for moisture and nutrients, that growth may be killed at seeding time or soon thereafter by undercutting with sweeps. Tillage in dry areas must make maximum use of scanty rainfall. The lister (double-mold board) plow, or middlebreaker, is here used to make water-impounding ridges that promote infiltration. The special problems of dryland farming will be considered.

Fertilizing and Conditioning of the Soil

Soil fertility is the quality of a soil that enables it to provide compounds in adequate amounts and proper balance to promote growth of plants when other factors (such as light, moisture, temperature, and soil structure) are favourable. Where fertility of a soil is not good, natural or manufactured materials may be added to supply the needed plant nutrients; these are called fertilizers, although the term is generally applied to largely inorganic materials other than lime or gypsum. Fertilizer grade is a conventional expression that indicates the percentage of plant nutrients in a fertilizer; thus, a 10–20–10 grade contains 10 percent nitrogen, 20 percent phosphoric oxide, and 10 percent potash. The green plant, however, requires more nutrients than these.

Essential Plant Nutrients

In total, the plant has need of at least 16 elements, of which the most important are carbon, hydrogen, oxygen, nitrogen, phosphorus, sulfur, potassium, calcium, and magnesium. The plant obtains carbon and hydrogen dioxide from the atmosphere; other nutrients are taken up from the soil. Although the plant contains sodium, iodine, and cobalt, these are apparently not essential. This is also true of silicon and aluminum.

Overall chemical analyses indicate that the total supply of nutrients in soils is usually high in comparison with the requirements of crop plants. Much of this potential supply, however, is bound tightly in forms that are not released to crops fast enough to give satisfactory growth. Because of this, the farmer is interested in measuring the available nutrient supply as contrasted to the total quantities.

The solid content of soils is broadly classified as organic and inorganic. Materials of organic origin range from fresh plant tissue to the more or less stable black or brown degradation product (humus) formed by biological decay. The organic matter is a potential source of nitrogen, phosphorus, and sulfur; it contains more than 95 percent of the total nitrogen, 5 to 60 percent of the total phosphorus, and 10 to 80 percent of the total sulfur. These three elements are cycled through the entire environment of living things (the biosphere). The soil organic matter can be considered as one of the storage points in these cycles. Where nonlegumes are grown in the absence of fertilizer or manures, the crop must gain its nitrogen supply from the organic matter; only a part, however, of the needed phosphorus and sulfur is so supplied.

The inorganic or mineral fraction, which comprises the bulk of most soils, is derived from rocks and their degradation products. The power to supply plant nutrients is much greater in the larger particles, sand and silt, than in the fine particles, or clay. The minerals that comprise the sand and silt in soil contain most of the elements essential for plant growth as a part of their structure. The difficulty is that these minerals decompose so slowly in soil that the rate of supply of the nutrient elements is usually insufficient for good growth of plants.

When the available supply of a given nutrient becomes depleted, its absence becomes a limiting factor in plant growth, and the addition of this nutrient to the soil will increase yields of dry matter. Excessive quantities of some nutrients may cause decrease in yield, however.

Determining Nutrient Needs

Determination of a crop's nutrient needs is an essential aspect of fertilizer technology. The appearance of a growing crop may indicate need of fertilizer; in some plants, however, the need for more or different nutrients may not be easily observable. If such a problem exists, its nature must be diagnosed, the degree of deficiency must be determined, and the amount and kind of fertilizer needed for a given yield must be found. There is no substitute for detailed examination of plants and soil conditions in the field, followed by simple fertilizer tests, quick tests of plant tissues, and analysis of soils and plants.

Sometimes plants show symptoms of poor nutrition. Chlorosis (general yellow or pale-green colour), for example, indicates lack of sulfur and nitrogen. Iron deficiency produces white or pale-yellow tissue. Symptoms can be misinterpreted, however. Plant disease can produce appearances resembling mineral deficiency, as can various organisms. Drought or improper cultivation or fertilizer application each may create deficiency symptoms.

After field diagnosis, the conclusions may be confirmed by experiments in a greenhouse or by making strip tests in the field. In strip tests, the fertilizer elements suspected of being deficient are added, singly or in combination, and the resulting plant growth observed. Next, it is necessary to determine the extent of the deficiency.

An experiment in the field can be conducted by adding nutrients to the crop at various rates. The resulting response of yield in relation to amount of nutrient supplied will indicate the supplying power of the unfertilized soil in terms of bushels or tons of produce. If the increase in yield is large, this practice will show that the soil has too little of a given nutrient. Such field experiments may not be practical, because they can cost too much in time and money. Soil-testing laboratories are available in most areas; they conduct chemical soil tests to estimate the availability of nutrients. Commercial soil-testing kits give results that may be very inaccurate, depending on techniques and interpretation. Actually, the most accurate system consists of laboratory analysis of the nutrient content of plant parts, such as the leaf. The results, when correlated with yield response to fertilizer application in field experiments, can give the best estimate of deficiency. Further development of remote sensing techniques, such as infrared photography, are under study and may ultimately become the most valuable technique for such estimates.

The Economics of Fertilizers

The practical goal is to determine how much nutrient material to add. Since the farmer wants to know how much profit to expect if he buys extra fertilizer, the tests are interpreted as an estimation of increased crop production that will result from nutrient additions. Cost of nutrients must be balanced against value of crop or even against alternative procedures, such as investing the money in something else with a greater potential return. The law of diminishing returns is well exemplified in fertilizer technology. Past a certain point, equal inputs of chemicals produce less and less yield increase. The goal of the farmer is to use fertilizer in such a way that the most profitable rate is employed.

Fertilizers can aid in making profitable changes in farming. Operators can reduce costs per unit of production and increase the margin of return over total cost by increasing rates of application of fertilizer on principal cash and feed crops. They are then in a position to invest in soil conservation and other improvements that are needed when shifting acreage from surplus crops to other uses.

Farm Manure

Among sources of organic matter and plant nutrients, farm manure has been of major importance in past years. Manure is understood to mean the refuse from stables and barnyards, including both excreta and straw or other bedding material, while the term fertilizer refers to chemicals. Large amounts of manure are produced by livestock; such manure has value in maintaining and improving soil because of the plant nutrients, humus, and organic substances contained in it.

As manure must be managed carefully in order to derive the most benefit from it, some farmers may be unwilling to expend the necessary time and effort. Manure must be carefully stored to minimize loss of nutrients, particularly nitrogen. It must be applied to the right kind of crop at the proper time. Also, additional fertilizer may be needed, such as phosphoric oxide, in order to gain full value of the nitrogen and potash that are contained in manure.

Manure is fertilizer graded as approximately 0.5–0.25–0.5 (percentages of nitrogen, phosphoric oxide, and potash), with at least two-thirds of the nitrogen in slow-acting forms. Commercial fertilizer equivalent to one ton (900 kilograms) of average manure can be purchased at a fairly low price. Furthermore, the expense of applying 100 pounds (45 kilograms) of 10–5–10 fertilizer is much less than the cost of applying 20 times as much manure. On properly tilled soils, the returns from fertilizer usually will be greater than from an equivalent amount of manure. The application of manure to a crop cannot be controlled as readily as can granulated fertilizer. In general, manure does not provide all the plant nutrients needed and fails to provide any that cannot be supplied by artificial fertilizers. Thus, there is a tendency to discount the value of manure as fertilizer. In underdeveloped countries, however, where artificial fertilizer may be costly or unavailable and where labour is relatively cheap, manure is attractive as a fertilizer.

The main benefits of manure are indirect. It supplies humus, which improves the soil's physical character by increasing its capacity to absorb and store water, by enhancement of aeration, and by favouring the activities of lower organisms. Manure incorporated into the topsoil will help prevent erosion from heavy rain and slow down evaporation of water from the surface. In effect, the value of manure as a mulching material may be greater than is its value as a source of essential plant nutrients.

Green Manuring

In reasonably humid areas, the practice of green manuring can improve yield and soil qualities. A green-manure crop is grown and plowed under for its beneficial effects, although during its growth it may be grazed. These green crops are usually annuals, either grasses or legumes, whose roots bear nodule bacteria capable of fixing atmospheric nitrogen. Among the advantages of green-manure crops are the addition of nitrogen to the soil; increase in general fertility level, reduction of erosion, improvement of physical condition, and reduction of nutrient loss from leaching. Disadvantages include the chance of not obtaining a satisfactory growth; the possibility that the cost of growing the manure crop may exceed the cost of applying commercial nitrogen; possible increases in disease, insects, and nematodes (parasitic worms); and possible exhaustion of soil moisture by the crop.

Green-manure crops are usually planted in the fall and turned under in the spring before the summer crop is sown. Their value as a source of nitrogen, particularly that of the legumes, is unquestioned for certain crops such as potatoes, cotton, and corn (maize); for other crops, such as peanuts (groundnuts; themselves legumes), the practice is questionable. Farmers are gradually turning away from growing green-manure crops except where the crop may also serve as winter cover for the land.

Compost, Peat and Sludge

Compost, peat, and sludge are used in agriculture and gardening as soil amendments rather than as fertilizers, because they have a low content of plant nutrients. They may be incorporated into the soil or mulched on the surface. Heavy rates of application are common.

Compost, or synthetic manure, is basically a mass of rotted organic matter made from waste-plant residues. Addition of nitrogen during decomposition is usually advisable. The result is a crumbly material that when added to soil does not compete with the crop for nitrogen. When properly prepared, it is free of obnoxious odours. Composts commonly contain about 2 percent nitrogen, 0.5 to 1 percent phosphorus, and about 2 percent potassium; if phosphate and potash are added while composting, those values are higher. The nitrogen of compost becomes available slowly and never approaches that available from inorganic sources. This slow release of nitrogen reduces leaching and extends availability over the whole growing season. Composts are essentially fertilizers with low nutrient content, which explains why large amounts are applied. The maximum benefits of composts on soil structure (better aggregation, pore spacing, and water storage) and on crop yield usually occur after several years of use.

In practical farming, the use of composted plant residues must be compared to the use of fresh residues. More beneficial soil effects usually accrue with less labour by simply turning under fresh residues; also, since one-half the organic matter is lost in composting, fresh residues applied at the same rate will cover twice the area that composted residues would cover. In areas where commercial fertilizers are expensive, labour is cheap, and implements are simple, however, composting meets the needs and is a logical practice.

Peat, composed of prehistoric plant remains that have accumulated under airless conditions in bogs, is a widely used organic soil amendment. Peat moss, the remains of sphagnum plants, is probably its most common form; it contains less than 1 percent nitrogen, with phosphorus and potassium below 0.1 percent. It is highly acid, with pH between 3 and 4.5 (a pH value of 7 is neutral and one above 7 basic). Peat improves the water-storage capability of soils and gives better structure to fine soils. Heavy applications of peat are usually the practice. It is used mostly by specialty-crop producers and on lawns and gardens.

Sewage sludge is the solid material remaining from the treatment of sewage. Its value for soil improvement depends on the method used for treating the sewage. Activated sludge, which results from aerobic (oxygen) treatment, contains 5 to 6 percent nitrogen and 1 to 3.5 percent of phosphorus. After suitable processing, it is sold as fertilizer and soil amendment for use on lawns, parks, and golf courses. It is rarely used in farming.

Liming

Liming to reduce soil acidity is practiced extensively in humid areas where rainfall leaches calcium and magnesium from the soil, thus creating an acid condition. Calcium and magnesium are major plant nutrients supplied by liming materials. Ground limestone is widely used for this purpose; its active agent, calcium carbonate, reacts with the soil to reduce its acidity. The calcium is then available for plant use. The typical limestones, especially dolomitic, contain magnesium carbonate as well, thus also supplying magnesium to the plant.

Another liming material is basic slag, a by-product of steel manufacture; its active ingredient is calcium silicate. Marl and chalk are soft, impure forms of limestone and are sometimes used as liming materials, as are oyster shells. Calcium sulfate (gypsum) and calcium chloride, however, are unsuitable for liming, for, although their calcium is readily soluble, they leave behind a residue that is harmful.

Lime is applied by mixing it uniformly with the surface layer of the soil. It may be applied at any time of the year on land plowed for spring crops or winter grain or on permanent pasture. After application, plowing, disking, or harrowing will mix it with the soil. Such tillage is usually necessary, because calcium migrates slowly downward in most soils. Lime is usually applied by trucks specially equipped and owned by custom operators.

Methods of Application

Fertilizers may be added to soil in solid, liquid, or gaseous forms, the choice depending on many factors. Generally, the farmer tries to obtain satisfactory yield at minimum cost in money and labour.

Manure can be applied as a liquid or a solid. When accumulated as a liquid from livestock areas, it may be stored in tanks until needed and then pumped into a distributing machine or into a sprinkler irrigation system. The method reduces labour, but the noxious odours are objectionable. The solid-manure spreader conveys the material to the field, shreds it, and spreads it uniformly over the land. The process can be carried out during convenient times, including winter, but rarely when the crop is growing.

Application of granulated or pelleted solid fertilizer has been aided by improved equipment design. Such devices, depending on design, can deposit fertilizer at the time of planting, side-dress a growing crop, or broadcast the material. Fertilizer attachments are available for most tractor-mounted planters and cultivators and for grain drills and some types of plows. They deposit fertilizer with the seed when planted, without damage to the seed, yet the nutrient is readily available during early growth. Placement of the fertilizer varies according to the types of crops; some crops require banding above the seed, while others are more successful when the fertilizer band is below the seed.

The use of liquid and ammonia fertilizers is growing, particularly of anhydrous ammonia, which is handled as a liquid under pressure but changes to gas when released to atmospheric pressure. Anhydrous ammonia, however, is highly corrosive, inflammable, and rather dangerous if not handled properly; thus, application equipment is quite specialized. Typically, the applicator is a chisel-shaped blade with a pipe mounted on its rear side to conduct the ammonia five to six inches (13 to 15 centimetres) below the surface. Pipes are fed from a pressure tank mounted above. Mixed liquid fertilizers containing nitrogen, phosphorus, and potassium may be applied directly to the surface—by field sprayers where close-growing crops are raised. Large areas can be covered rapidly by use of aircraft, which can distribute both liquid and dry fertilizer.

The Future for Fertilizers

Future trends in fertilizer technology may be predicted by extrapolating from current developments. Mixtures and materials with high percentages of plant nutrients will dominate the field. Better ways of providing nitrogen, the most expensive of the three major nutrients, will be forthcoming, including increased use of anhydrous ammonia, ammonium nitrate, and urea. Nonleachable nitrogen, for example, can be obtained through the urea–formaldehyde (ureaform) reaction, and ammonium metaphosphate offers a concentrated liquid product. Micronutrients, or trace elements, specific to particular geographical areas will come into increasing use, as will custom mixing and bulk selling of mixtures containing several nutrients based on reliable soil and plant data.

"Complete environment" seeding in which seed, fertilizer, and water are incorporated in a biodegradable (decomposable in the soil) tape may come into use; with the tape planted, no further fertilizer or water will be needed until growth is well established. Such techniques using biodegradable tapes have already been developed on a small scale for use by home gardeners. Finally, larger and more precise fertilizing machines will be developed and adopted.

Factors in Cropping

Cropping System

The kind and sequence of crops grown over a period of time on a given area of soil can be described as the cropping system. It may be a pattern of regular rotation of different crops or one of growing only one crop year after year on the same area.

Crop Rotation

Early agricultural experiments showed the value of crop rotations that included a legume sod crop in the regular sequence. Such a system generally maintains productivity, aids in keeping soil structure favourable, and tends to reduce erosion. Alfalfa, sweet clover, red clover, and Ladino clover are considered effective for building up nitrogen. Some legumes, however, do not leave nitrogen behind in the soil because it is deposited as protein in the harvested seed; soybeans are an example. Turning under the top growth of a legume aids in adding nitrogen. Though yields of grains are higher when they are rotated with legumes, it is difficult to determine how much of the improvement depends on the nitrogen added by the legume and how much on improved soil structure or fewer insects and disease.

The determination of the best rotation depends upon whether the crops compete with each other (*i.e.*, if growing one crop lowers the yield of its successor) or complement each other; and the output of one crop on a given acreage leads to increased output of the other. This desirable complementary relationship exists only when one crop or soil-management practice concurrent with it provides nutrient or conditions required by the other crop. In this circumstance, grasses and legumes may complement grains or row crops by furnishing nitrogen, controlling erosion and pests, and improving soil structure to such an extent that greater production is achieved. The reverse can also occur; in certain prairie soils, continuous growing of deep-rooted legumes depletes soil moisture, and subsequent forage yield is improved by frequent plowing of the sod and planting of corn. In high-rainfall or irrigated areas, forage stands deteriorate from winter killing, disease, or grazing, to a point where a year of grain in the rotation allows an improved stand of forage later. Fallow (idle) land is complementary to wheat and other small grains in subhumid areas such as the Great Plains of the United States; such rotation is quite beneficial to wheat yield. Complementary relationships between crops can be terminated by the application of the physical law of diminishing returns, however, and give way to competition.

Both long-range and short-range profits motivate the farmer as cropping systems are examined in relationship to soil erosion. Excessive loss of soil to streams, rivers, and reservoirs is unacceptable to public policy as well as economically damaging to the farmer, and crop rotations that promote erosion are minimized. Soil losses are least from fields in continuous sod and most from continuous row crops. If row crops are grown in rotation with sod, the erosive susceptibility of row crops is reduced over a period of time. Peanuts (groundnuts), potatoes, tobacco, cotton, sugar beets, and some vegetables, and similar row crops that require frequent cultivation (intertillage) and leave minimal post-harvest residue are most likely to permit serious erosion. Less erosive are row crops such as corn (maize), sugarcane, and grain sorghum, which require less cultivation and leave more residues. Small grains such as wheat, oats, barley, and rye usually permit less erosion than the row crops. Among sod crops, grasses or grass–legume mixtures are less erosive than pure stands of legumes such as alfalfa. Fortunately, cropping systems that tend to control soil erosion usually tend also to give better yields than systems that promote excessive erosion. These results from increased availability of water to the plants and increased amounts of nutrients, which in erosive systems are washed away and lost.

Monoculture

The practice of growing the same crop each year on a given acreage, monoculture, has not been generally successful in the past, because nonlegume crops usually exhaust the nitrogen in the soil, with a resulting reduction in yields; this is particularly true in humid regions. The advent of low-cost nitrogen fertilizers, however, has induced reconsideration of the possible advantages of monoculture. These advantages can best be discussed in terms of a hypothetical general farm where it may be desirable to produce several different kinds of crops: the question to be answered is whether monoculture can do better than rotational systems in producing these crops while still maintaining productivity.

Advantages of Monoculture

First, if different kinds of soil exist on the farm, a monoculture system may permit each crop to be grown on the soil best suited to it. Forage crops, for example, could be confined to steep land to minimize erosion; intertilled crops could be planted on the better soils with gentle slopes. Wet areas could be used continuously for crops not requiring early-spring field operations, while dry soils could be used for drought-resistant crops such as sorghums or winter small grains.

Second, the fertility level of the soil can be adjusted to fit one crop more precisely than it can be adjusted to fit all the crops in a rotation.

Third, where continuous cropping is practiced and perennial forage crops are used, regular reseedings are avoided. This is an advantage, because each seeding is accompanied by the possibility of failure.

Fourth, systems based on monoculture usually offer greater flexibility in planning the system to meet year to year changes in the need for various crops. Part of the acreage can be shifted from one crop to another without upsetting the total farm cropping plan.

Disadvantages of Monoculture

On the other hand, requirements for successful monoculture are more demanding of management skill than are sod-based rotations. The entire nitrogen need of nonlegume crops must be met by purchased fertilizers or by use of manure. Closer attention to soil erosion is necessary, except for perennial sod. Soil-structure problems can become severe where intertilled crops are grown continuously. In monoculture, the farmer is completely dependent on chemical insecticides, disease-resistant plant varieties, soil fumigation, and similar methods of controlling insects and diseases that are usually controlled by crop rotation.

Thus, the choices of cropping systems that maintain good productivity, minimize soil losses, and are in harmony with demand and desired business organization are not easily made. The growing use of systems analysis will undoubtedly aid in rational selection among the bewildering array of possibilities.

Crop Protection

Crops are vulnerable to attack, damage, and competition. Insects, plant disease, nematodes, rodents, weeds, and air pollution are among the many enemies that can reduce crop yields and deny man the use of some of his farm-stored crops.

Insects, for example, can destroy a crop in a relatively short time. Control measures for many years have engaged the attention of farmer and scientist, yet full success has not been achieved, and the battle continues. The problem is further complicated by the fact that control measures not only kill unwanted insects, but also may harm honey bees as well as the parasites and predators that destroy insect pests.

At least 10,000 species of insects are classed as unwanted. Of these, several hundred species are particularly destructive and require some degree of control. They destroy food as well as the forage, pasture, and grain needed to produce livestock; and, in addition, they carry and transmit many diseases of plants and animals.

Chemical Control of Insects

Insecticides generally are effective, cheap, and safe if handled correctly; the good derived from them, however, can be partly offset by adverse effects. Chlorinated hydrocarbon insecticides such as DDT, for example, may leave residues toxic to beneficial insects, fish, and other wildlife; the insecticides may be found in meat and milk, or they may persist in the soil. Another problem is that some species of insects build up resistance to chlorinated hydrocarbon, organic phosphate, and carbamate insecticides.

These disadvantages can be overcome only by persistent search for new and safer insecticides accompanied by wide use of nonchemical insect control.

A wide range of organophosphate and carbamate materials is now available. These can be applied to avoid most of the problems related to residues. Malathion and carbaryl, for example, are used to control insects in areas where persistent materials might appear later in meat or milk and can also be applied in areas where fish and wildlife might be affected. Those two chemicals offer a broad range of toxicity to insect pests. Unlike chlorinated hydrocarbons, they can be applied up to within a day or so of harvest without harm to many crops; they are dangerous, however, to those who apply them and must be handled with care.

Some insecticides are effective in very small amounts. This fact has stimulated development of ultralow-volume technology, where special equipment permits dispersal of low volumes of undiluted chemicals, which offers cost advantages as well as drastic reduction of the chemicals in the environment. For example, six to 16 ounces (170 to 450 grams) per acre of Malathion may be effective against grasshoppers, boll weevil, cereal-leaf beetle (*Oulema melanopus*), mosquitoes, and the beet leafhopper (*Circulifer tenellus*). Formulation of chemicals in granules rather than sprays offers some advantages in use and applications; among others, it reduces the amount needed and also lessens the chance of adverse effects on beneficial insects and wildlife.

Certain insects that attack cotton, vegetables, and forage crops may be controlled by chemicals absorbed by the plant. Called systemics, they are placed with the seed at planting time. The chemical is taken up by the plant, and insects die when they attempt to feed on the leaf or stem. Beneficial insects that do not feed on the plant remain unharmed.

Non-Chemical Control of Insects

Mechanical and Cultural Controls

Light traps that give off radiation that attracts insects have been under test for many years. They have been somewhat successful in controlling the codling moth (*Carpocapsa pomonella*) and the tobacco hornworm (*Protoparce sexta*).

Use of reflective aluminum strips, placed like a mulch in vegetable fields, has reduced or prevented aphid attack and thus protected cucumbers, squash, and watermelons from mosaic diseases. This technique may supplant insecticides, which frequently do not kill aphids quickly enough to prevent crop losses from virus transmitted by them.

For stored products, heat or cold can control many insects that frequent such places. Also, changing the proportions of oxygen, nitrogen, and carbon dioxide in the storage atmosphere can provide control.

It has been discovered that, if adult Indian-meal moths (*Plodia interpunctella*) were exposed to certain wavelengths of sound during the egg-laying period, their reproduction was reduced by 75 percent. The sound waves had a similar effect on flour beetles (*Tribolium* species). Further development is needed, but this method offers potential as a nonchemical control. Other types of physical energy can also kill insects. Light waves, high-frequency electric fields, high-intensity radio frequencies, and gamma radiation have been investigated; some offer promise.

Certain cultural practices can prevent or reduce insect crop damage. These include destruction of crop residues, deep plowing, crop rotation, use of fertilizers, strip-cropping, irrigation, and scheduled planting operations. Such practices are useful but cannot be relied upon entirely to eliminate severe infestations.

Biological Controls

The question of using biological controls has always been of considerable public interest. The control agents include parasites, predators, diseases, protozoa, and nematodes that attack the insect pests. Biological controls cannot replace insecticides entirely, because nature provides for survival of both beneficial and destructive insects. Before the population of a parasite or predator can expand, a high population of the host species must also be present. Sometimes the control agents are far outnumbered by the pest insect. Parasites and predators have furnished good control of the Japanese beetle (*Popillia japonica*), European corn borer (*Pyrausta nubilalis*), alfalfa aphid (*Therioaphis maculata*), alfalfa weevil (*Hypera postica*), and several others.

Microbial agents can be used for control. There exist about 1,100 viruses, bacteria, fungi, protozoa, rickettsiae, and nematodes that parasitize insects. Many pathogens are specific to a particular insect but are harmless to man and domestic animals. It is a possibility that insect pathogens can be produced, packaged, distributed, and applied in much the same way as insecticides.

The ideal solution to insect-control problems is to plant crop varieties that are resistant to attack. The only difficulty is that such varieties are not universally available, and development entails a very long process.

Sterilization of male insects by gamma radiation and their release into a population of wild insects is a promising approach. It has proved successful in control of screwworms and fruit flies, replacing chemicals in some areas. Chemical attractants, which lure insects into contact with small amounts of insecticide or a sterilant, also offer much promise.

All aspects of insect control considered, it is possible that "integrated control," coordinated employment of more than one method, may be the answer. Combining resistant varieties with a systemic insecticide that leaves the parasites and predators unharmed, for example, has been successful in combatting the spotted alfalfa aphid in California.

Preliminary reduction of heavy infestation by chemical spray combined with bait, followed by the sterile-insect technique, provides another example of integrated control. Use of sex attractant in light traps, plus special management of postharvest residues, has controlled the tobacco hornworm. Other examples might be cited, but the principal value of such control methods lies in using less insecticide and thus contributing to maintenance of a good environment.

Control of Plant Diseases and Nematodes

Insects, of course, are not the only agents hazardous to crops. Plant diseases and the microscopic worms called nematodes have the potential of creating wholesale destruction of crops, especially those grown in regions of wide weather fluctuation. In fact, these plant pests sometimes limit the kinds and varieties of crops that can be grown. The damage they cause may sometimes be mistaken for that caused by unfavourable weather. Epidemics may destroy crops completely.

As with insects, control of plant diseases and nematodes covers a broad spectrum of measures: use of chemicals, resistant varieties, quarantine, forecasting and warning, cultural practices, heat treatment, and others. Furthermore, most plant virus diseases are transmitted by insect carriers, so control of insects is linked to control of disease.

Nematodes and plant disease can at times be controlled fairly well by crop rotation, deep plowing, and burning of stubble and debris that remain after harvest. Though burning destroys aboveground organisms and permits economical control by chemicals, it contributes to air pollution and destroys organic matter. In another technique, propane-gas flame is applied to living plants as well as stubble to kill disease spores. A virus disease of sugarcane is controlled by heating diseased cuttings in hot-air ovens. Stem rot disease of peanuts (groundnuts) can be controlled by plowing under dead plant debris or by planting the seed on a raised bed followed by application of a preemergence weed killer.

Successful control of epidemic plant disease may depend on prompt application of chemicals before the disease outbreak. Many governments operate plant-disease forecasting and warning services for farmers. The service is based primarily on analysis of temperature, rainfall, humidity, and dew—all factors that can create conditions favourable to disease outbreaks.

Weed Control

Weed control is vital to agriculture, because weeds decrease yields, increase production costs, interfere with harvest, and lower product quality. Weeds also impede irrigation water-flow, interfere with pesticide application, and harbour disease organisms.

Early methods of weed control included mowing, flooding, cultivating, smothering, burning, and crop rotation. Though these methods are still important, other means are perhaps more typical today, particularly the use of herbicide (plant-killing) chemicals. Another technique is to introduce insects that attack only the unwanted plant and destroy it while leaving the crop plants unharmed.

The inadequacy of the cultural, mechanical, and biological control systems, however, stimulated the rapid development of chemical usage since World War II. Herbicides have had an impact on crop production, changing many cultural and mechanical agricultural operations.

Herbicides are formulated as wettable powders, granular materials, emulsions, and solutions. Any of them may be applied as a spot treatment, broadcast, placed in bands, or put directly on a specific plant part. When formulated as solutions or emulsions, the chemical is mixed with water or oil.

Spraying is the most common method, permitting extremely small amounts to be applied uniformly because of dilution. Sprays can be accurately directed underneath growing plants, and calibration and rate control are easier with spray machines than with granular applicators. Granular formulations have advantages under some conditions, however. The use of herbicides must be integrated into the overall farm program because the optimum date and application rate depend on the crop stage, the weed stage, weather conditions, and other factors.

Careful use of herbicides in farm production lowers cost, resulting in a more economical product for the consumer. Herbicides cut the costs of raising cotton, for example, by reducing labour requirements for weed control up to 60 percent. Herbicides replace hand labour in growing crops, labour that is no longer available in developed nations at costs the farmer can afford. Machines for chemical application are widely available.

When used as directed, herbicides are generally safe, not only for the operator but also for wildlife and livestock. The greatest difficulty lies in accidental injury to crop plants resulting from drift and from residues in the soil, particularly if residues enter water courses.

The future of chemical pesticides and herbicides is under debate by those who manufacture, sell, and use them and by those who are concerned about environmental quality. The value of an assured food and fibre supply at reasonable cost is undeniable, and chemicals contribute much toward this. These substances also cause undesirable effects upon the environment, however, and indeed can be toxic to a wide range of organisms. This fact will demand an increasing amount of care in using chemicals, perhaps enforced by law, along with increasing use of nonchemical control techniques.

Harvesting and Crop Processing

Harvesting Machinery

Harvesting machinery is generally classified by crop: reapers for cutting cereal grains and threshers for separating the seed from the plant. The more modern combine cuts, threshes, and cleans the grain in one operation. Corn (maize) harvesting is performed by mechanical corn pickers that snap the ears from the stalk so that only the grain and cobs are harvested. Corn shelling may be done mechanically in the field, after or with picking. Stripper-type cotton harvesters, which strip the entire plant of both open and unopened bolls, work best late in the season after frost has killed the green vegetative growth. Hay and forage machines include mowers, crushers, windrowers, field choppers, balers, and some machines that press the hay into wafers or pellets.

Grass, legumes, corn (maize), and other crops are often put into silos to keep them in a succulent and fermented state rather than stored dry as hay. To make silage, the crops must be cut up to permit tight packing in the silo, producing anaerobic fermentation and preventing formation of mold. Almost all silage crops are cut in the field with a forage harvester that cuts and chops the crop immediately or picks up and chops a windrow that has been cut and raked earlier.

Root crops are harvested with diggers and digger-pickers, which often pull up clods, stones, and vines with the crop. Though some machines carry workers who manually sort out extraneous material, this task is increasingly being performed mechanically. Modern sugar-beet harvesters lift the whole root from the ground, clean the earth from it, and deliver it to a bin or wagon. Sometimes the beet tops are removed before harvest of the roots and used for cattle feed. Peanuts (groundnuts) are lifted, vines and all, and allowed to dry before removal of the pods.

Tobacco-harvesting aids may be classified in three principal ways, according to the harvesting and curing methods used, which depend on the type of tobacco and its use. Flue-cured tobacco, a large plant that may stand three to four feet (90 to 120 centimetres) high, is harvested with machines that carry several workers who ride the lower platforms of the machines, cut the leaves, and place them on conveyor belts, where the leaves are tied mechanically or by hand. Burley tobacco has usually been harvested by workers using a machete-type knife. After cutting, the large end of the stalk is fixed onto the sharpened end of a stick, which—when loaded with a number of stalks—is hung by hand in a tobacco barn for curing. Researchers are attempting to mechanize the cutting, impaling, and hanging of burley tobacco. Little has been done, however, toward the mechanization of the harvesting of the small aromatic tobacco leaves, which are grown in the shade, picked by hand, tied with a string, and then hung for curing.

Tree-crop harvesting is accomplished by hand or with mechanical shakers. Vegetable crops such as asparagus, lettuce, and cabbage are still harvested largely by hand, though scarcity and high cost of field labour has led to some mechanization in this area, notably with tomatoes.

Crop-processing Machinery

Machinery is widely used to prepare crops for convenient transportation, for safe storage, for the market, and for feeding to livestock. Advances in such machines have been rapid, particularly with new crops, increased yields, multiple-crop practices, and changing techniques.

In the most common method of crop drying, the crop, usually grain, is spread on floors or mats and stirred frequently while exposed to the sun. Such systems, though extremely common in the underdeveloped countries, are very slow and dependent on the weather. Forced-air-drying systems allow the farmer much more freedom in choosing grain varieties and harvest time. Fairly simple in operation, these systems have been gaining popularity in the tropics. Heat is often added to increase air temperatures during the drying period.

In a process called dryeration, wet corn (maize) is placed in a batch or continuous dryer. After losing 10 to 12 percent of its moisture, the hot corn is transferred to the dryeration cooling bin, in which it is tempered for six to 10 hours and then slowly cooled by ventilation for 10 hours. This process reduces kernel damage and increases dryer output.

High moisture in stored hay not only causes rapid deterioration of its value as feed but often results in spontaneous combustion. When hay is first cut, it usually contains 70 percent or more moisture. It wilts and quickly dries to a moisture content of about 40 percent. At this stage, it can be dried to a safe storage condition, about 15 percent moisture, by blowing air through it, sometimes with supplemental heat.

Feed-processing mills, often referred to as feed grinders, are used principally for milling cereals for livestock feed, which aids digestion. The ground material is usually fairly coarse and at times may only be crushed. Modern mills frequently are designed to allow the farmer to grind the grain and to mix in various other ingredients in desired quantities.

Other types of crop-processing machinery include machines that separate desirable seed from weed seed, stems and leaves, and dirt; grading machinery to classify seed by width, length, or thickness; fruit graders and separators; and cotton gins, which separate cotton seeds from the fibres.

Regional Variations in Techniques

Dryland Farming

Dryland farming refers to production of crops without irrigation in regions where annual precipitation is less than 20 inches (500 millimetres). Where rainfall is less than 15 inches (400 millimetres) per year, winter wheat is the most favoured crop, although spring wheat is planted in some areas where severe winter killing may occur. (Grain sorghum is another crop grown in these areas.) Where some summer rainfall occurs, dry beans are an important crop. All dryland crop yield is mainly dependent on precipitation, but practices of soil management exert great influence on moisture availability and nutrient supply.

Where rainfall exceeds 15 inches (380 millimetres), the variety of crop possibilities is increased. In areas of favourable soils and moisture, seed alfalfa is grown, as is barley. Some grass seed may be grown particularly crested wheat grass of various types.

Fallow System and Tillage Techniques

Dryland farming is made possible mainly by the fallow system of farming, a practice dating from ancient times. Basically, the term fallow refers to land that is plowed and tilled but left unseeded during a growing season. The practice of alternating wheat and fallow assumes that by clean cultivation the moisture received during the fallow period is stored for use during the crop season. Available soil nitrogen increases and weeds are controlled during the fallow period. One risk lies in the exposure of soil while fallow, leaving it susceptible to wind and water erosion. Modern power machinery has tended to reduce this risk.

Procedures and kinds of tillage that are comparatively new have proved effective in controlling erosion and improving water intake. Moldboard and disk plows are being replaced with chisels, sweeps, and other tools that stir and loosen the soil but leave the straw on the surface. Where the amount of straw or residue remaining from the previous crop is not excessive, this trashy fallow system works well, and tillage implements are designed to increase its effectiveness.

Contour tillage helps to prevent excessive runoff on moderate slopes. Broad terraces can aid in such moisture conservation. Steeper slopes are planted to permanent cover. Compacted zones at a depth of five to eight inches (13 to 20 centimetres) can be caused by tillage. As such zones interfere with storage of moisture, they can be controlled by growing deep-rooted alfalfa at intervals, or the compacted zone can be broken by fall tillage with chisels or sweeps set to a depth just below the zone of compaction. Such deep tillage will result in reduced runoff and deeper moisture penetration.

When using power machinery in dryland farming, the timing of operations is important. The soil is broken in the fall or early spring before weeds or volunteer grain can deplete the moisture. Use of a rod weeder or similar equipment during fallow can control the weeds. Planting is timed to occur during the short period in fall or spring when temperature and moisture are favourable.

Fertilizer Use

Fertilizer is an important component of dryland technology. For example, 20 pounds per acre (22 kilograms per hectare) of nitrogen are recommended where rainfall is less than 13 inches (330 millimetres), ranging up to 60 pounds per acre (67 kilograms per hectare) where more rain is available; those figures refer to the production of wheat, but they are applicable to other dryland-farming areas. Where average annual precipitation is less than 12 inches (300 millimetres), the use of nitrogen is limited to years where moisture outlook is exceptionally favourable. Nitrogen fertilizer can be applied either in fall or spring. Band placement or broadcast techniques are utilized. Good results are obtained from broadcast spring application of nitrate fertilizer, and fall application of ammonia has also been successful. Local climates and rainfall patterns also determine choice of fertilizer and time of application.

Crops and Planting Methods

Alfalfa grown for seed on drylands is planted in rows, usually two to three feet (60 to 90 centimetres) apart; cultivation between rows is required during the first year. Alfalfa is also grown for forage where favourable. This practice builds nitrogen and organic matter, while improving soil structure. These legumes can be rotated with wheat if rain is between 16 and 18 inches (400 and 450 millimetres) and will increase the yield of wheat.

Contour farming; strip cropping.

Other crops, such as cotton, peanuts (groundnuts), and grain sorghum, can be grown successfully in dryland agriculture. Where those crops are produced on sandy soils, special techniques are necessary to reduce soil blowing and drifting. Cotton and peanuts do not produce sufficient crop residues for protection from wind erosion, while sorghum does. For this reason, many farmers in such areas use various row combinations of cotton or peanuts with grain sorghum; two rows of grain sorghum and four to eight rows of peanuts in alternating strips is a popular technique. Another is to use a two-year rotation of cotton and grain sorghum, in which two rows are cropped and two rows are fallow. These systems not only afford protection from wind erosion but also promote effective use of soil moisture.

Tropical Farming

The area of the world bounded roughly on the north by the Tropic of Cancer and on the south by the Tropic of Capricorn, a vast land that embraces large parts of Latin America, Africa, India, Australia, and Southeast Asia, contains climates less favourable to agriculture and human settlement than those of the temperate zones. Within this Equator-centred area occur the climates known as tropical, which are characterized by two general types: warm and wet, and warm with partly deficient rainfall. In either, the total precipitation is usually quite heavy, which leaches the tropical soils of nutrients. The area also has high temperatures with little variation the year round. The combination of high temperature and high rainfall causes organic matter to decompose quickly, leaving the soil deficient in humus. Vegetation flourishes in the tropics, along with weeds, insects, and disease organisms. Important climatic variations occur, depending upon land elevation.

Tropical crops include coconut, palm oil, rice, sugar, pineapple, sisal, cocoa, tea, coffee, jute, rubber, pepper, banana, and many others. In certain highland tropical areas, however, the crops common to temperate-climate agriculture can also be grown. The amount of tropical land well-suited to agriculture, however, is limited.

Plant Pest Problems

The abundance of plant pests in the tropics, including weeds and disease, makes agriculture successful mainly in the plantation system, where needed control measures can be financed. The alternative is to move from deteriorated land to newer fields; this practice of shifting agriculture has also been common, because tropical soils lose their productive capacity so rapidly. The practice probably cannot be continued indefinitely, however, because of increasing population pressure.

The largest quantities of commercial tropical products originate in plantations, where skilled management is combined with sufficient capital to provide mechanized equipment. This is particularly true in the production of coffee, cocoa, rubber, coconut, banana, pineapple, sugarcane, and others. Much rice is produced in the Asian tropics and Indonesia, however, on small farms with intensive hand labour and simple tools, where the prime mover is likely to be the ox or the water buffalo, not the tractor.

Water Management

Drainage, irrigation, and other special techniques of water management are important in tropical agriculture. An example is the cultivation of rice and sugarcane in the fertile coastal areas of Guyana. Originally through private enterprise and later by government efforts, large coastal areas were "empoldered" (diked) to keep back the sea in front and floods from the rivers in the rear. With a mean annual rainfall of 90 inches (2,300 millimetres), drainage is a critical factor; in fact, the system cannot discharge all possible floodwater, and so the crops must tolerate occasional drowning. With gravity drainage effective only at low tide, the drainage gates are opened on the ebbing tide and closed on the rising tide. Great difficulty is encountered in keeping the outlets unclogged by the heavy sediment discharge. Since rain does not always fall when it is needed, many fields are irrigated. Most of the rice soil is specially tilled after plowing in order to create a better seedbed under the water, using tractors operating in water four to six inches

(10 to 15 centimetres) deep. After this special tillage, the seeds are broadcast in one to two inches (2.5 to five centimetres) of water. Though maintenance and operation of such an intricate water-control system are not simple, Guyana rice production has been doubled through its use.

Mechanical Problems

Mechanization faces many obstacles before wide adoption is possible in tropical regions. Difficult soils, stones, stumps, abundant labour, resistance from farmers, lack of incentives, lack of skills, lack of capital, low wages, high cost of machines, lack of dealer service, fragmented land ownership, all contribute to slow development of mechanization. Tropical soils differ markedly from those in the countries that manufacture land-preparation machinery, making adaptation of new design necessary. The encountering of stones, wood, trash, and termite mounds causes machines to break down. Depressing climatic conditions reduce the performance of the machine operators. Tropical farm regions are notoriously irregular or mountainous, impeding intensive machine culture. The best soils in Brazil require special erosion controls, reducing the potential for large-scale mechanization. One of the greatest overall impediments to mechanization is the fear that unemployment might result from it, a failure to understand that economic development and higher living standards depend partly on increasing the productivity of labour.

As an example of the problems encountered in mechanizing tropical crops, the harvesting experience of a large sugarcane plantation in Trinidad is illuminating. On flatland of some 30,000 acres (12,000 hectares), the cane is grown on heavy clay soil in a climate with 50 inches (1,300 millimetres) of rain during the seven-month wet season and 10 inches during the five-month dry season. By 1960 the rising wage rate made harvest mechanization imperative. First, the traditional "bed" system, which functioned to remove floodwater, was changed to ridge planting; this made it possible for machines to operate and was a remarkable change in itself. Then it was decided to harvest with the cane combine, which tops, cuts, chops, and loads the chopped cane into transport vehicles. Although the combine is complicated and requires considerable power, it was deemed better than mechanical half-measures.

By 1969 the combines, however, were harvesting only 12.8 percent of the flatland crop, indicating that mechanization was far from complete. Three factors were responsible: first, the cane combines required extensive maintenance plus very expensive replacement parts. Second, it was difficult to mobilize a transport system to receive the output of the combines with any degree of economy. Third, the social problem of displaced workers had to be considered. The combines increased labour productivity sixfold over hand harvesting; thus, their introduction had to be slowed until surplus workers could be accommodated elsewhere. The limited success of this mechanization project indicates how complicated such a process really is.

Taking the largest view of possibilities for improving tropical agriculture, the most promising inputs of technology are improved crop varieties and increased use of fertilizers.

Other Specialized Techniques

Hydroponics

The term hydroponics denotes soilless culture of plants. The possibilities of this technique have received considerable attention in recent years. In hydroponics, an outgrowth of laboratory techniques long used by scientists, plants are grown with their roots immersed in a water solution containing necessary minerals or rooted in a sand medium kept moistened by such a solution. Soilless culture of plants is similar in principle but larger in scale. A typical hydroponics technique has plants supported in a bed of peat, wood fibre, or similar material, on a wire screen with the roots dipping into the solution below. Aeration of the solution is provided. In another method, the plants are rooted in a medium of sand or gravel contained in a shallow tank into which the solution is pumped at intervals by automatic control. Between pumpings, the solution drains slowly down into a reservoir tank. Hydroponic techniques are practiced on a small scale both out-of-doors and in greenhouses.

Of the elements known to be necessary for plant growth, carbon, oxygen, and hydrogen are obtained by the plant from atmospheric gases or from soil water. The others are all obtained as mineral salts from the soil. The elements absorbed as salts—iron, manganese, boron, copper, zinc, and molybdenum—are required in minute quantities and are called the micronutrients. The principal elements that must be provided as dissolved salts in hydroponic techniques are nitrogen, phosphorus, sulfur, potassium, calcium, and magnesium. Numerous solutions have been devised to fulfill these requirements.

Crop yields of some plants can be obtained fully equal to those obtained on fertile soils. Wide-scale crop production by hydroponics, however, would be economic only for certain intensive types of agriculture or under special conditions. Some greenhouse crops, both vegetables and flowers, are grown by this method. In regions having no soil or extemely infertile soil but with favourable climate, hydroponic techniques have been very useful; for example, on some of the coral islands of the Pacific.

Greenhouses

The greenhouse is typically a structure whose roof and sides are transparent or translucent, permitting a sufficient quality and quantity of solar radiation to enter the structure for photosynthesis. It allows the growing of crops independently of the outside climate, since its interior temperature and humidity can be controlled. Greenhouses vary in size and complexity from small home or hobby structures to large commercial units covering an acre or more of land. An even smaller greenhouse might be termed the hot bed, a glass-topped box containing fermenting organic matter; the fermentation process yields heat, allowing the gardener to start plants from seed in early spring for later transplanting.

The basic construction of a greenhouse consists of a light but sturdy frame capable of resisting winds and other loads. Conventional foundations usually support vertical walls; the roof may be gabled, trussed, or arched. The conventional greenhouse is fitted with glass panes, but plastic-film or fibre-glass panels often supplant glass.

Maintenance of temperature within the greenhouse is difficult because of fluctuating outside conditions. When the sun shines brightly, little heat is needed, and the heating system must be controlled in some way to prevent injury to the crop. Hot water, steam, electric cable, or warm-air furnaces provide the heat, which is usually controlled by thermostat. Temperatures in greenhouses are regulated to suit the crop. Typical ranges are from 40 °F (4 °C) for lettuce, violets, carnations, and sweet peas to 70 °F (21 °C) for cucumbers, tomatoes, and orchids.

Cooling is often required during summer days in warm climates. Ventilation is the simplest technique, reducing inside temperature to near that of the outdoors. Additional cooling by refrigeration may be required; in dry regions, the evaporative cooler is efficient and also increases the relative humidity within the structure. Another form of environmental control consists of adding extra carbon dioxide to the air if the crop requires it for extra photosynthetic efficiency.

The commercial-greenhouse operator usually grows vegetables or ornamental plants. Such production makes more demands on the grower, because he must assume many of the tasks normally handled by nature in the open fields. He must regulate the temperature, ventilate, adjust the amount of entering sunlight, provide soil moisture, fertilize, and even facilitate pollination. During the off-season, the structure must be cleaned and fumigated, its soil restructured, and mechanical equipment checked. Mechanization of greenhouse operations has lagged far behind the pattern of agriculture in general. Disease is a particularly serious hazard in greenhouse farming, requiring constant attention and use of chemicals.

Agricultural Biotechnology

Agricultural biotechnology is the use of different scientific techniques to modify plants and animals. The undesirable characteristics like susceptibility to diseases and low productivity are bred out. If there is a particular trait that the plant or animal can benefit from, it can be bred in by using a gene that contains the characteristic.

Biotechnology has especially been beneficial in improving agricultural productivity and increasing the resistance of plants to diseases. Scientists do this by studying the DNA.

They first identify the gene that would be beneficial to the plant or animal then work with the characteristics conferred in a precise and exact manner to achieve the desired outcome.

Advantages of Agricultural Biotechnology

Biotechnology has been beneficial in many ways. First, stabilized plants that have higher yields have been produced successfully. The resistance of these plants to pests, diseases and abiotic factors such as rainfall has played a major role in increasing the yields.

Animal feeds are being improved by biotechnology to increase their nutrient intake and reduce environmental wastes. Another advantage of biotechnology is that it has led to the development of better vaccines that don't necessarily have to be stored in very cold temperatures. Penicillin, one of the most important components of antibiotics was produced through biotechnology.

Uses of Agricultural Biotechnology

Genetic Engineering

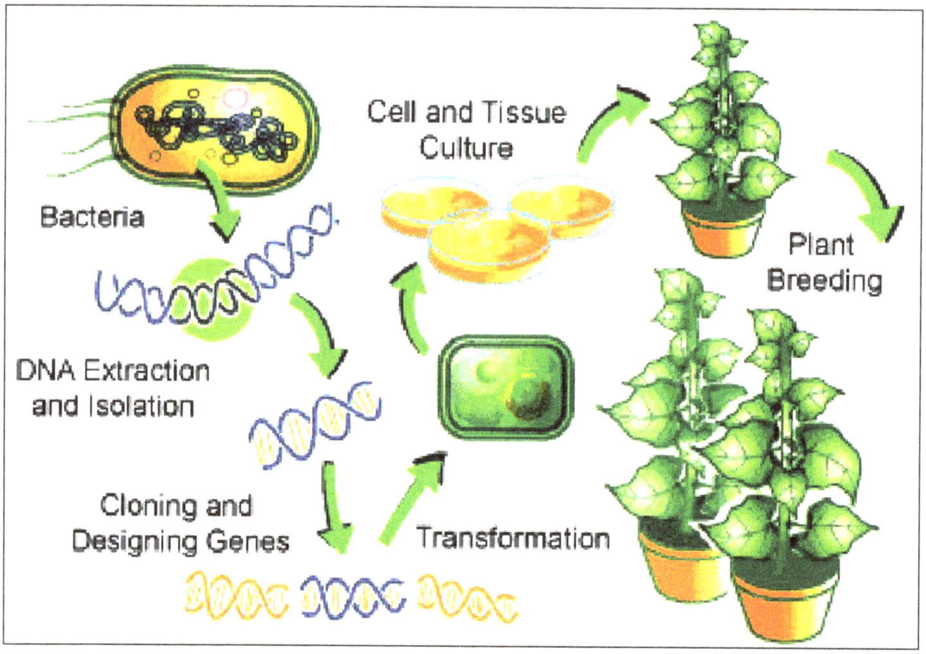

Genetic engineering, also referred to as genetic improvement or modification, is the movement of a gene from one organism to another. This process allows for the transfer of a useful characteristic into an organism by inserting it with a gene containing the particular trait. In crops, genetic engineering has been used to increase productivity and resistance to weeds and harsh weather conditions.

Molecular Markers

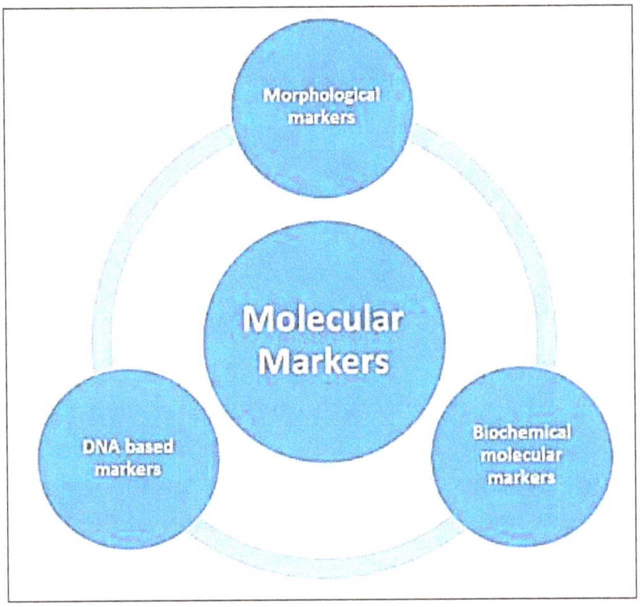

Breeding was previously concerned with the removal and insertion of desirable physical traits, an example being the aggression of Bulldogs. By studying the DNA, scientists were able to find molecular markers that showed traits that were not visible. Using molecular markers, breeding has been made more precise and accurate, and this has countered the undesirable characteristics that may have appeared in future generations.

Vaccines

Biotechnology is used for making vaccines for both animals and human beings. These vaccines are better than the traditional ones because they are cheaper, safer, and can survive warmer tropical temperatures. Vaccines to prevent new infections have also been developed using biotechnology.

Genomics

A genome is an entire set of chromosomes found in the DNA, and through the study of genomes and genetic mechanisms, breakthroughs have been made in biotechnology. Through genomics, the structure, function, location, and impact of a particular gene and genome are identified. This makes it easy to determine the characteristics that will be transferred to another organism and the exact results of the transfer of the gene.

Tissue Culture

This technique is used to produce a plant that is free from undesirable characteristics, which are mostly, diseases. A disease-free plant part is used to generate types that are disease-free. The different types of plants in which tissue culture works include bananas, avocados, mangoes, coffee, and papaya, among others.

Biotechnology in Agriculture and Food Production

In 1990, the first food produced as a result of biotechnology was released into the market. Ever since 1995, there has been a steady production of genetically modified foods all over the world, and by 2003, 7 million farmers, mostly in developing countries, were planting biotech crops.

Biotechnology has positively influenced the economy and social life of most developing countries. The increased food production by biotech plants means that more people can now enjoy food security while spending less on pesticides. This has, subsequently, led to increased standards of living.

Biotechnology in agriculture, both for plants and for animals is a reality. It offers a tool through which these organisms can be understood, and their genetic resource management improved. Plants and animals have, thus been able to increase their productivity and get better resistance to diseases through the study of their genes and the manipulation of their characteristics. There are different ways through which biotechnology is done, and they include genetic engineering, vaccination, molecular marker, tissue culture, and genomics. Technology is evolving and so does the world of science, and who knows, there might be another fascinating breakthrough in genetics and biotechnology.

References

- Agricultural-engineering: sarvgyan.com, Retrieved 14, August 2020
- Animal-sciences, agricultural-sciences-311781: britannica.com, Retrieved 20, February 2020
- Agricultural-technology: britannica.com, Retrieved 10, July 2020
- What-is-agricultural-biotechnology: mixerdirect.com, Retrieved 28, March 2020

Chapter 3

Genetic Modification in Plant and Crops

Genetically modified plants are the plants that are engineered to create enhanced crops, new colors and to deliver vaccines. Some of the techniques that are used to create genetically modified plants are protoplast isolation, somatic cell hybridization, gene gun method, and electroporation. The topics elaborated in this chapter will help in gaining a better perspective about these techniques.

Genetically modified (GM) plants are also called transgenic plants. They are designed to acquire useful quality attributes such as insect resistance, herbicide tolerance, abiotic stress tolerance, disease resistance, high nutritional quality, high yield potential, delayed ripening, enhanced ornamental value, male sterility, and production of edible vaccines. Another major goal for raising the GM plants is their application as bioreactors for the production of nutraceuticals, therapeutic agents, antigens, monoclonal antibody fragments biopolymers, and so forth. Thus, GM plants can potentially affect many aspects of modern society, including agricultural production and medical treatment. Despite these potential applications, the use of GM plants for human welfare has been restricted owing to various concerns raised by the public and the critics. These concerns are divided into different categories, namely, health, nutritional, environmental, ecological, socioeconomic, and ethical concerns. These concerns include those arising due to properties of GM plants themselves, those resulting from the spread of the transgenes to other organisms, and also those resulting from their release into the environment. Such concerns have led to the withdrawal of commercialization of Bt cotton and Bt brinjal in India.

The campaign against GM plants was fueled by the instances of transgenic potatoes reported to be deleterious to rats, contamination of commercial corn products with unapproved StarLink and killing of monarch butterfly by Bt corn pollen. Furthermore, the nongovernmental organizations (NGOs) such as Gene Campaign, Center for Sustainable Agriculture, Research Foundation for Science, Technology and Ecology, Greenpeace, and Friends of the Earth have also raised concerns related to genetic manipulation of plants. The regulators, activists, media personnel and scientific journals have been undiscriminating and overly tolerant of the misrepresentations and distortions of anti-GM activists. There are not even scientific explanations for some of the concerns, but today the amount of misinformation is such that it has become difficult to separate truth from public perception about the GM plants. The biotechnology scientists, however, believe that GM plants should be given public acceptance because most of the concerns are not specific for GM plants and can exist for non-GM plants as well.

Concerns Related to Health and Nutritional Status

In case the products of GM plants are to be consumed by humans and animals, there is always a fear and risk in the society that these plants may create health problems or may lead to the development of newer microbial strains that may be pathogenic. Further, the plants themselves may be susceptible to such risks. Public and critics are also skeptical about the nutritional content and quality of the GM plants.

Susceptibility to Allergens

One of the major distressing problems with nontraditional proteins in GM foods is the risk of introducing allergens (usually glycoproteins) into the food supply of humans and animals. The public is concerned about the nature of these new food proteins as their allergenic or nonallergenic qualities are unknown. Allergenicity has been demonstrated in transgenic soybeans due to the transfer of a major food allergen from Brazil nuts. On the other hand, the scientists believe that the food allergens are found only in a few defined sources (peanut and other grain legumes, shellfish, tree nuts, etc.), and hence, only a dozen foods may produce allergic reactions. Moreover, allergenicity occurs when these food allergens are present in large proportions in the food and the individuals are sensitized to them over time to cause any adverse effects. Thus, it is highly unlikely for new allergens to be introduced into the food supply from GM plants.

Transfer of Antibiotic Resistance Gene to Microbes and Reduced Efficacy of Antibiotic Therapy

Public is also concerned about the potential risks associated with gene transfer from plants to microbes. It is speculated that the consumption of GM foods containing antibiotic resistance marker gene (e.g., Npt II gene encoding neomycin phosphotransferase for resistance to kanamycin and neomycin or gene encoding β-lactamase for resistance to ampicillin) by humans and animals may lead to transfer of these genes from GM food to microflora in the gut of humans and animals or to the pathogens in the environment transforming them into strains that are resistant to antibiotic therapy.

The transfer of antibiotic resistance gene to unrelated microorganisms such as Aspergillus niger has also been demonstrated. Biotechnology scientists, however, are of the opinion that the Npt II gene used to develop GM plants currently in the market is safe for use because there is no evidence of allergenicity or toxicity related to it. Moreover, humans are also susceptible to consuming several kanamycin resistant bacteria that occur naturally in the environment. Human gut is reported to contain 1012 kanamycin-resistant bacteria and by consuming a tomato harboring Npt II gene, the increase in frequency of kanamycin-resistant bacteria in the gut amounts only to 10–6%. Furthermore, acid conditions prevalent in stomach or rumen inactivate or degrade the encoded enzyme, neomycin phosphotransferase II.

Also neomycin phosphotransferase II requires ATP for its activity, which is present in extremely low concentrations in the gut. Regarding the use of gene for selection of bacterial recombinants, it is not transferred to plants. Moreover, the gene is considered safe because it does not encode for any product in plants. The Npt II and genes have been declared safe to use in GM plants. The public is, however, reluctant to accept this fact. Looking to the views of public, scientists have also developed nonresistance based selectable marker genes such as green fluorescent protein encoding gene (Gfp) and β-glucuronidase gene (Uid A). Besides, intron-containing Npt II gene has also been assessed as an efficient selectable marker in plant transformation. Due to insertion of intron in the Npt II gene, the theoretical risk of gene flow from GM plants to enteric bacteria is eliminated. Strategies for the removal of antibiotic resistance genes have also been devised.

One such strategy is the cloning of selectable marker gene and the transgene on two separate transfer DNA (T-DNA) molecules in a single plasmid or on two separate plasmids that are contained in one or more Agrobacterium tumefaciens strains used for plant transformation. The transgene and selectable marker gene are, thus, inserted at the loci, which should recombine at reasonably high frequencies so that the transgene can be segregated from the selectable marker gene in the next generation. Second strategy to eliminate the selectable marker gene is to flank it with direct repeats of recognition sites for a site-specific recombinase so that the marker gene can be easily excised from the plant genome by recombinase-mediated site-specific recombination. Examples included in this category are the Cre/lox recombination system of bacteriophage P1, Flp/frt recombination system of yeast 2 μm plasmid and R/Rs system of Zygosaccharomyces rouxii. A common feature of these systems is that the first round of transformation produces transgenic plants with the selection marker between two directly oriented recognition sites for the respective recombinase. After expression of recombinase, either by crossing in plants expressing the enzyme, by transient expression via second transformation, or by the use of an inducible promoter, the recombinase reaction is initiated resulting in marker-free transgenic plants. Marker gene may also be eliminated by placing it on a transposable element resulting in its loss after transposition.

The transgene by itself may be mobile and the activation of transposase allows the relocation of the desired transgene to a new chromosomal position. Genetic crosses and/or segregation may dissociate the two transgenes. Another novel strategy for the production of marker-free GM plants involves DNA deletion based on intrachromosomal homologous recombination between two homologous sequences, for example, by incorporating att sequence of λ bacteriophage.

Development of New-Line Microbial Strains

The third health risk is related to the ability of GM plants to create new toxic organisms. It is speculated that some nonpest microbial strains may acquire pathogenic trait by gene flow from GM plants. The risk can also be a new host being infected by a virus or recombining to form a more deadly virulent virus. Some plant pathologists also hypothesize that development of virus-resistant plants may allow viruses to infect new hosts through transencapsidation. Virus-resistant plants may also lead to the creation of new viruses through an exchange of genetic material or recombination between RNA virus genomes. Another matter of concern is that a small fraction of the DNA released from GM plants into soil may bind to the clay particles and hence protected from degradation. It is speculated that the soil bacteria may undergo transformation with the exogenous DNA of GM plant. This is, however, a rare possibility as the amount of DNA derived from GM plants as a proportion of the total DNA in the soil is likely to be very small, even if such plants are grown on a commercial scale. Moreover, the longevity of DNA in soil depends on various factors, including soil type and the presence of deoxyribonucleases in soil. target DNA from transgenic tobacco plants gets degraded within 40 days.

Skepticism About Nutritional Status

Critics of GM crops have raised various concerns about the potential of golden rice to combat vitamin A deficiency (VAD). The primary concern amongst these is the presence of insufficient vitamin A in golden rice. There are still doubts about the speed of degradation of vitamin A after harvesting the plant and the amount of vitamin A left after cooking. Vandana Shiva, an Indian anti-GMO activist, has criticized golden rice by saying "the golden rice is a hoax," "golden rice is a blind approach for blindness control," and "golden rice is just a recipe for creating hunger and malnutrition". She argues that the golden rice fails to pass the vitamin A need test and is incapable of removing VAD. It is calculated that one serving contains 30 g of rice on dry weight basis and golden rice can provide only 9.9 µg of vitamin A, that is, only 1.32% of the required daily allowance (RDA) of 750 µg. Even with the daily consumption of 100 g golden rice, only 4.4% of the required daily allowance will be met. Thus, an adult has to consume 2 kg 272 g of golden rice per day to complete his daily requirements of vitamin A. She is also of the view that, besides creating VAD, golden rice will also create deficiency in other micronutrients and nutrients.

This is because the raw milled rice has a low content of fat (0.5 g/100 g), which is necessary for vitamin A uptake, low content of protein (6.8 g/100 g), which is required as a carrier molecule, and low content of iron (0.7 g/100 g), which is required for the conversion of β-carotene to vitamin A. Friends of the Earth, Greenpeace, and Vandana Shiva further emphasize that there is no need of golden rice to combat VAD as superior alternatives such as sweet potato, green leafy vegetables, coriander, amaranth, carrot, pumpkin, mango, jackfruit exist in nature. It is reported that certain underutilized plants also exhibit far more nutritional value (vitamin A and other nutrients) than golden rice, for example, a combination of rice and leaves of Moringa (drumstick) tree, a native to India. Similarly, in contrast to rice, amaranth grain contains forty times more calcium, four times iron, and twice as much protein.

The ragi millet, grown in India, has thirty five times more calcium than rice, twice as much iron, and five times more minerals. It is opined that golden rice is not capable of increasing the production of β-carotene. Even if the target of 33.3 µg of vitamin A in 100 g of rice is achieved, it will be only 2.8% of β-carotene that can be obtained from amaranth leaves, 2.4% as that obtained from coriander leaves, curry leaves, and drumstick leaves. Thus, a far more efficient route to removing VAD is biodiversity conservation and propagation of naturally occurring vitamin A rich plants (wild-type or underutilized) in agriculture and diets. Even the World Bank has admitted that rediscovering and use of local plants and conservation of vitamin A rich green leafy vegetables and fruits have dramatically reduced VAD threatened children over the past 20 years in very cheap and efficient ways.

It is also speculated that the cultivation of golden rice will lead to major water scarcity since it is a water intensive crop and displaces water prudent sources of vitamin A. The scientists, on the other hand, believe that the traditional breeding methods have been unsuccessful in producing crops containing a high vitamin A concentration and most national authorities rely on expensive and complicated supplementation programs to address the problem. They also believe that a varied diet is beyond the means of many of the poor and they have to rely on one or few foods to provide complete nutrition, for example, rice.

Thus, golden rice may be a useful tool to help treat the problem of VAD in young children living in the tropics. They also emphasize that the critics are ignoring the fact that VAD disorders result from a deficiency of vitamin A and not its complete absence in the diet and the VAD individuals lack only 10%–50% of their daily requirements. Hence, any additional contribution toward daily requirements would be useful. In 2005, a team of researchers at Syngenta have produced a variety of golden rice, called "Golden rice 2," which produces twenty-three times more carotenoids than golden rice (up to 37µg/g) and preferentially accumulates β-carotene (up to 31µg/g of the 37µg/g of carotenoids). The Rockefeller Foundation emphasized that the new strains of golden rice contain substantially higher levels of β-carotene than the early versions on which the opponents based their calculations. In order to meet the RDA, 144g of the most high-yielding golden rice strains would have to be eaten.

Environmental and Ecological Concerns

Large-scale cultivation of GM plants expressing viral and bacterial genes and their release into the environment is considered to be a threat and called as "genetic pollution" by the critics. The risk of a transgene spreading in the environment is related to the likelihood for out-crossing, horizontal gene transfer, and the phenotype imparted by the gene. Debates about the commercial introduction of GM plants in some parts of the world have led to questions about their potential impact on the environment unless necessary safeguards are taken into account. It should, however, be acknowledged that agriculture inevitably has an impact on the environment and these concerns are not specific for GM plants.

Transgene Escape to Wild-Type Plants

There is a potential risk that the GM plants may hybridize (or cross-breed) with sexually compatible wild-type species. This genetic exchange is possible due to wind pollination, biotic pollination or seed dispersal. This may have an impact on the environment through the production of hybrids and their progeny. In an example, virus-resistant squash commercialized in 1994 was demonstrated to transfer its virus resistance gene to wild squash (Cucurbita pepo), an agricultural weed native to the southern United States, thereby decreasing its value to squash breeders. On the other hand, it is significant to note that for an effective pollen transfer to occur, the GM plants must be close enough to the wild species, should flower at same time, and must be genetically compatible. Further, the risk of any gene transfer to related weedy species through pollen has been eliminated by devising chloroplast transformation procedures. This is because, in many crop species, chloroplasts display only maternal inheritance.

Selective Advantage to GM Plants in Natural Environments and Generation of Super weeds

The concern of gene flow from GM plants to weedy relatives via pollination is quite intense. It is considered that the transfer of encoded characteristics to weed species could potentially give them a selective advantage, consequently leading to the generation of "superweeds." Moreover, the newly introduced traits may make a plant, especially herbicide tolerant plant, more persistent or invasive (weedy) in agricultural habitats. It is, however, pertinent to note that the risk of gene transfer to weeds is similar with both conventional and GM plants and is not contingent on how these genes have been introduced into plants. Such a risk of gene flow has always existed since the advent of modern plant breeding, even when there were no GM plants, and this can occur where possible. Several studies have demonstrated that tolerance to particular herbicide is often more likely to develop by evolution from within the weed gene pool rather than by gene flow from herbicide-tolerant plants. Nevertheless, the current scientific evidence indicates that the weediness arises from many different characters and that the addition of one gene is unlikely to cause a crop to become a weed. The transfer of novel genes from transgenics (or even conventionally bred plants) to weeds depends on the nature of the novel gene and the biology and ecology of the recipient weed species. The probability of successful out-crossing thus depends on sexual compatibility, physical proximity, distance of pollen movement both out of and into the GM plants, and ecology of recipient species. Thus, only a few plants such as oilseed rape, barley, wheat, beans, and sugar beet can hybridize with weeds. For example, oilseed rape has been reported to hybridize with hoary mustard, wild radish, and other wild Brassica species. Furthermore, the transfer of herbicide tolerance gene is unlikely to confer any competitive advantage to hybrids outside agricultural areas. It is also comforting to recognize that there is no proven evidence of enhanced persistence or invasiveness of GM plants and no major super weeds have developed so far.

Effect on Nutritional Composition of Plants

It is also speculated that the nutritional composition of GM products may be affected in GM plants. Another concern is that the transgenes from animals (obtained from fishes, mouse, human, and microbes) introduced into GM plant for molecular farming may pose a risk of changing the fundamental nature of vegetables. In a study, it was reported that as compared to non-GM soybean, GM plants exhibited lower levels of isoflavones . This finding also raised a doubt on the regulatory system for the release of the GM plants. However, later it was found that the concentration of isoflavone in GM soybean was within the normal range.

Mixing Genes from Unrelated Species (Interbreeding)

The public is worried about the risk that the GM plants can spread through nature and interbreed with natural organisms, thereby contaminating "non-GM" environments. This would in turn affect the future generations in an unforeseeable and uncontrollable way . Such worries, however, ignore the history of plant breeding and the existing overwhelming sequence similarity of genes across kingdoms.

Development of Tolerance to Target Herbicide

It is viewed that the repeated use of the same herbicide in the same area to remove weeds amongst genetically modified herbicide-resistant crops (HRCs) (tolerant to single herbicide) will exacerbate the problem of herbicide-tolerant weeds . Another matter of concern relates to the plants carrying different herbicide tolerance genes to become multiply tolerant to several herbicides by pollination between adjacent plants . In several closely studied examples in Canada, farmers have detected oilseed rape plants tolerant to three different herbicides (note that two were acquired from GM plants and the third possibly from conventional breeding) . The development of multiple tolerances in "volunteer" crop plants (from seeds remaining viable in agricultural soil) may also exert an impact on the environment by necessitating the use of less environment-friendly and possibly outdated herbicides by the farmers. On the other hand, the proponents believe that herbicide resistance develops due to excessive application of herbicide and is not exclusively associated with gene transfer from genetically modified HRCs. Thus, the pressure on weeds to evolve resistant biotypes has been reported to be pronounced with the excessive application of herbicides such as glyphosate, sulphonylureas, and imidazolinones.

Sustainable Resistance in Insect Pests

It is possible that the widespread use of disease-resistant GM plants may lead to the evolution of several insect pests that are resistant to pesticides. For example, Bt crops may develop resistance to Bt biopesticide, a permitted biopesticide successfully used by organic farmers in the integrated pest management (IPM) programs. There is to date no reported evidence of insect resistance to Bt crops under field conditions although Bt resistant insects (e.g., cotton budworm and bollworm) have been observed in areas where Bt biopesticides are sprayed on crops . It has been a matter of concern that the development of such resistance may lead to the loss of the potential of the Bt biopesticide, which may in turn make it necessary for organic farmers to resort to less environmentally acceptable chemical pesticides. Therefore, proper resistance management strategies along with this comparatively newer technology are imperative. The most widely used is the 'high-dose refuge' strategy designed to prevent or delay the emergence of Bt toxin-resistant insects. Scientists are of the opinion that this strategy should be followed without fail, as the rate of noncompliance can increase the risk of plant resistance breakdown.

Harm to Nontarget Organisms

Nontarget effect, that is, undesirable effect of a novel gene (usually conferring pest or disease resistance) on "friendly" organisms in the environment, is another concern related to GM plants . As many nontarget microbes harbor on plant surfaces or some insects harbor on flowers, it becomes quite challenging to target the insect resistance gene product to appropriate plant tissues and hence kill pests without exerting any adverse effect on friendly organisms such as pollinators and biological control agents. This is particularly difficult where the benign or beneficial organism is related and physiologically similar to the pest to be targeted. One of the most significant studies of nontarget impacts of GM plants has been the killing of monarch butterfly in the United States by Bt insecticidal proteins. It should, however, be noted that the pesticidal sprays used on Bt or non-Bt corn may be more harmful to the monarch butterfly as compared to Bt corn pollen. Thus, in evaluating the use of Bt crops and the possible environmental damage caused, it is important to take into account the environmental damage caused by the use of pesticides in agriculture generally. It is argued that millions of birds and billions of insects, both harmful and beneficial, are killed each year due to excessive use of pesticides. It is, however, suggested that the scale and pattern of use may mitigate the effects of Bt on nontarget populations. Furthermore, when toxins are produced within plant tissues, nontarget organisms are exposed to a much lesser extent than with spray applications because only those organisms which feed on the plant tissues come into contact with the toxin.

Harmful effect of Bt toxin residues in the soil after harvest of the GM crop on soil invertebrates has been another matter of concern. An investigation of the effect of Cry1Ab released from the roots and crop residues on soil organisms revealed the presence of toxin in the guts and casts of tested earthworms. There was, however, no significant difference in their mortality or weight. Moreover, no difference in the total number of other soil organisms (including nematodes, protozoa, bacteria, and fungi) between the soil rhizosphere of Bt and non-Bt crops was detected.

Increased Use of Chemicals in Agriculture

On one hand, the transgenes conferring herbicide resistance have been criticized because these would maintain, if not promote, the use of herbicides and their attendant problems [125, 126]. Similarly, there is a concern that the insect-resistant and disease-resistant GM plants will increase the application of insecticides and pesticides, respectively. On the contrary, reports demonstrate that there is no significant change in the overall amount of herbicide use in the United States since the introduction of GM soybeans . An analysis by soybean growers at the United States has shown that $7.2 millions of other herbicides were replaced by $5.4 millions of glyphosate . This substitution, thus, resulted in the replacement of highly toxic and more persistent herbicides with that of glyphosate. Furthermore, it has been reported that herbicide-tolerant oilseed rape eliminates the use of >6,000 tons of herbicide in the growing season.

Loss of Biodiversity

The public has long been worried about the loss of plant biodiversity due to global industrialization, urbanization, and the popularity of conventionally-bred high-yielding varieties. It is speculated that the biodiversity will be further threatened due to the encouraging use of GM plants. This is because development of GM plants may favor monocultures, that is, plants of a single kind, which are best suitable for one or other conditions or produce one product. Further, the transformation of more natural ecosystems into agricultural lands for planting GM plants is adding to this ecological instability.

Another point of concern is the loss of weed diversity that may occur due to gene flow from HRCs to weeds . It is argued that because the currently available HRCs confer tolerance to broad-spectrum herbicides such as glufosinate and glyphosate, their extensive use may shift the diversity of weeds in agricultural habitats. However, weeds exhibit considerable plasticity and adapt to a wide range of cultivation practices. Experience with conventional agriculture has shown that weed species composition varies within the same crop among different fields and at different times of year. Thus, weed population shifts are natural ecological phenomena in crop management and should not be viewed as exclusive to GM plants.

Unpredictable Gene Expression

It is speculated that the random gene insertion, transgene instability, and genomic disruption due to gene transfer may result in unpredictable gene expression. Such a risk is, however, unlikely to be unique to GM plants or of any significance considering our current knowledge of genomic flux in plants.

Alteration in Evolutionary Pattern

Plants adapt to the fluctuations in the environment through changing their genes and developing better races called "evolved races." These mutations, however, occur at a very low frequency (i.e., one in about 109/gene/generation). It is hypothesized that the cultivation of GM plants by the farmers at an increasing rate throughout the world may change the evolutionary pattern drastically . Another concern is the evolution of non-GM plants through hybridization with GM plants.

Loss of Ecosystem in Marginal Lands

As new plants are introduced mainly to marginal lands, loss of natural ecosystems in these areas has also been a matter of concern.

Contamination of Soil and Water

It is also sometimes argued that the widespread introduction of HRCs will increase the use of herbicides, which will in turn contribute to the contamination of soil and ground water. However, this is not the case. The cultivation of HRCs in the United States has been reported to facilitate zero-till agronomic system, which contributes to a reduction in soil erosion. The release of Bt toxin into the soil after harvest of Bt crops is also viewed as a risk factor associated with the cultivation of Bt crops [123, 124, 130]. It has been found that Bttoxins remain active in soil; however, it is not necessarily an environmental hazard because Bt toxins must be ingested and affect only selected groups of insects. Moreover, the potential leaching rate of Bt toxin is reduced due to its binding and adsorption on clay particles.

Genetically Modified Crops

The term genetically modified (GM), as it is commonly used, refers to the transfer of genes between organisms using a series of laboratory techniques for cloning genes, splicing DNA segments together, and inserting genes into cells. Collectively, these techniques are known as recombinant DNA technology. Other terms used for GM plants or foods derived from them are genetically modified organism (GMO), genetically engineered (GE), bioengineered, and transgenic. 'Genetically modified' is an imprecise term and a potentially confusing one, in that virtually everything we eat has been modified genetically through domestication from wild species and many generations of selection by humans for desirable traits. The term is used here because it is the one most widely used to indicate the use of recombinant DNA technology.

Traits that have been Modified in GM Crops

Insect-resistant crops contain genes from the soil bacterium Bacillus thuringiensis (Bt). The protein produced in the plant by the Bt gene is toxic to a targeted group of insects–for example European corn borer or corn rootworm–but not to mammals. The most common herbicide tolerant (HT) crops are known as Roundup Ready, meaning they are tolerant to glyphosate (the active ingredient in Roundup herbicide). Glyphosate inactivates a key enzyme involved in amino acid synthesis that is present in all green plants; therefore, it is an effective broad spectrum herbicide against nearly all weeds. Roundup Ready crops have been engineered to produce a resistant form of the enzyme, so they remain healthy even after being sprayed with glyphosate. Some cultivars of corn and cotton are referred to as 'stacked', meaning they have transgenes for both insect resistance and HT.

Potential GM Crops of the Future

Some potential applications of GM crop technology are:

- Nutritional enhancement: Higher vitamin content; more healthful fatty acid profiles;
- Stress tolerance: Tolerance to high and low temperatures, salinity, and drought;
- Disease resistance: For example, orange trees resistant to citrus greening disease or American chestnut trees resistant to fungal blight;
- Biofuels: Plants with altered cell wall composition for more efficient conversion to ethanol;
- Phytoremediation: Plants that extract and concentrate contaminants like heavy metals from polluted sites.

Difference between GM Technology and Plant Breeding Techniques

The era of scientific crop improvement dates back to around 1900, when the impact of Gregor Mendel's studies on trait inheritance in peas became widely recognized. Since then, a broad range of techniques has been developed to improve crop yields, quality, and resistance to disease, insects, and environmental stress. Most plant breeding programs rely on manual cross-pollination between genetically distinct plants to create new combinations of genes. The progeny plants are intensively evaluated over several generations and the best ones are selected for potential release as new varieties. An example is a tomato variety that is selected for disease resistance and tolerance to cool temperatures. Other techniques included within the conventional plant breeding toolbox are development of hybrid varieties by crossing two parental strains to produce offspring with increased vigor; and induced mutations to create useful variation. GM technology is much more precise in that it transfers only the desired gene or genes to the recipient plant. Another branch of agricultural biotechnology—distinct from GM technology—involves selecting plants for DNA patterns known to be associated with favorable traits such as higher yield or disease resistance.

The Shared DNA Code

Most organisms store their genetic information in the form of DNA molecules in chromosomes. The sequence of chemical bases in a DNA strand encodes a specific order of amino acids, which are the building blocks of proteins. Proteins carry out many functions in cells and tissues, which together are responsible for an organism's characteristics. Because most life forms share this same language of heredity—and due to scientific advances in molecular biology—it is now possible to transfer a gene from one species to another, for example from a bacterium to a plant, and have it function in its new host.

Types of Genetically Modified Plants

Transgenic Plants

Transgenic plants or genetically modified plants are plants whose DNA is modified using genetic engineering techniques. In most cases the aim is to introduce a new trait to the plant which does not occur naturally in this species. Examples include resistance to certain pests, diseases or environmental conditions, or the production of a certain nutrient or pharmaceutical agent.

A transgenic crop plant contains a gene or genes which have been artificially inserted instead of the plant acquiring them through pollination. The inserted gene sequence, known as the transgene, may come from another unrelated plant or from a completely different species. Plants containing transgenes are often called genetically modified or GM crops.

Reasons for Making Transgenic Crop Plants

- The process of transgenic plant development primarily aims at assembling a combination of genes in a crop plant which will make it as useful and productive as possible. Depending on where and for what purpose the plant is grown, desirable genes may provide features such as higher yield or improved quality, pest or disease resistance, or tolerance to heat, cold and drought.
- Combining the best genes in one plant is a long and difficult process, especially as traditional plant breeding has been limited to artificially crossing plants within the same species or with closely related species to bring different genes together.

  For example, a gene for protein in soybean could not be transferred to a completely different crop such as corn using traditional techniques. Transgenic technology enables plant breeders to bring together in one plant useful genes from a wide range of living sources, not just from within the crop species or from closely related plants.
- This technology provides the means for identifying and isolating genes controlling specific characteristics in one kind of organism, and for moving copies of those genes into another quite different organism, which will then also have those characteristics.
- This powerful tool enables plant breeders to do what they have always done-to generate more useful and productive crop varieties containing new combinations of genes-but it expands the possibilities beyond the limitations imposed by traditional cross-pollination and selection techniques.

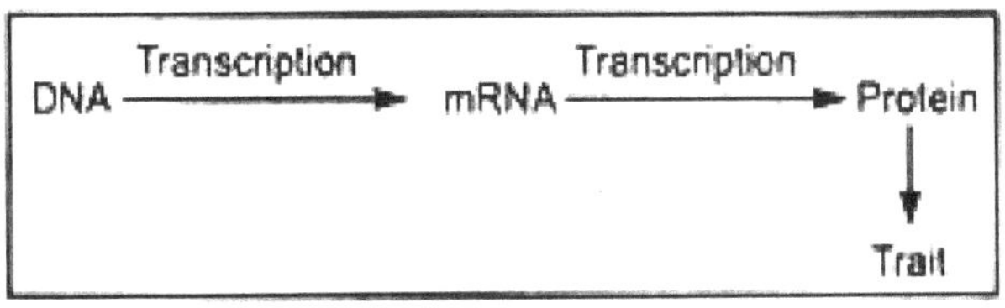

Fundamentals of Transgenic Plant Development

The underlying reason that transgenic plants can be constructed is the universal presence of DNA (deoxyribonucleic acid) in the cells of all living organisms. This molecule stores the organism's genetic information and orchestrates the metabolic processes of life. Genes are discrete segments of DNA that encode the information necessary for assembly of a specific protein.

A specific protein (or an enzyme) encodes for a particular trait. In the production of a transgenic plant our primary aim is to transfer a foreign gene, encoding for some novel traits, into the genome of the plant stably.

After transferring we also need the transgene to integrate and express in the plant's cells. This process as a whole generates a new variety of plant which is new in its own kind and interests us in its massive large scale cultivation.

Steps Involved in the Production of Transgenic Plants

The fundamental steps involved in the transgenic plant production are as follows:

Step: Identifying, Isolation and Cloning of Genes for Agriculturally Important Traits

The very first step in the generation of transgenic plant is to identify and isolate the novel transgene that we want to transfer into the genome of the target plant. Usually, identifying a single gene involved with a trait is not sufficient. We also have to understand how the gene is regulated, what other effects it might have on the plant, and how it interacts with other genes active in the same biochemical pathway.

Step: Designing Gene Construct for Insertion

After entering the plant cell the transgene must inter-grate into the genome of the plant stably and express itself successfully so as to produce higher amount of transgenic protein which will be indirectly reflected in the trait controlled by it.

To achieve this we have to design a "gene construct" or gene-set, having all the DNA segments necessary to achieve the integration and expression of the transgene. Once a gene has been isolated and cloned (amplified in a bacterial vector), it must undergo several modifications before it can be effectively inserted into a plant.

A gene-set which will be transferred to the target plant has following segments:

A Promoter Sequence

This must be added for the gene to be correctly expressed (i.e., translated into a protein product). The promoter is the on/off switch that controls when and where in the plant the gene will be expressed. To date, most promoters in transgenic crop varieties have been "constitutive", i.e., causing gene expression throughout the life cycle of the plant in most tissues.

The most commonly used constitutive promoter is CaMV 35S, from the cauliflower mosaic virus, which generally results in a high degree of expression in plants. Other promoters are more specific and respond to cues in the plant's internal or external environment. An example of a light-inducible promoteris the promoter from the cab gene, encoding the major chlorophyll a/b binding protein.

The Transgene

Sometimes, the transgene is modified to achieve greater expression in a plant.

For example, the Bt gene for insect resistance is of bacterial origin and has a higher percentage of A-T nucleotide pairs compared to plants, which prefer G-C nucleotide pairs. In a clever modification, researchers substituted A-T nucleotides with G-C nucleotides in the Bt gene without significantly changing the amino acid sequence. The result was enhanced production of the gene product in plant cells.

Termination Sequence

This signals to the cellular machinery that the end of the gene sequence has been reached.

A Selectable Marker Gene

This is added in order to identify plant cells or tissues that have successfully integrated the transgene. This is necessary because achieving incorporation and expression of transgenes in plant cells is a rare event, occurring in just a few per cent of the targeted tissues or cells.

Selectable marker genes encode proteins that provide resistance to agents that are normally toxic to plants, such as antibiotics or herbicides. Only plant cells that have integrated the selectable marker gene will survive when grown on a medium containing the appropriate antibiotic or herbicide.

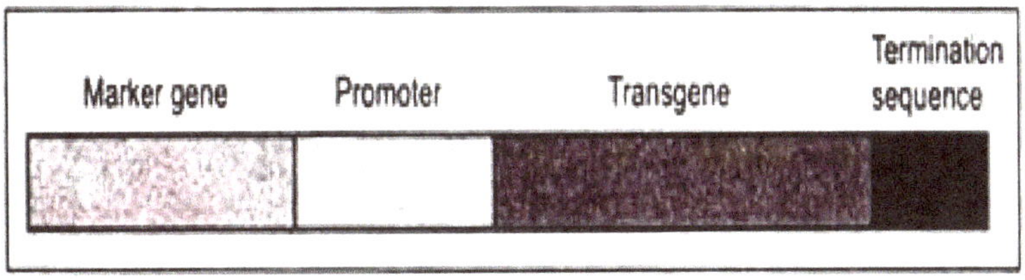

The essential features of an ideal reporter gene are:

1. Lack of endogenous activity in plant cells of the concerned enzyme,
2. An efficient and easy detection,
3. A relatively rapid degradation of the enzyme.

The commonly used selectable marker genes include those conferring resistance to the antibiotics kanamycin (nptII, encoding neomycin phosphotransferase) and hygromycin (hptIV, encoding hygromycin phosphotransferase, isolated from E. coli); and broad range herbicides glyphosate (modified versions of the enzyme EPSPS, 5-enolpyruvate shikimate-3-phosphate synthase, isolated from E. coli or Salmonella typhimurium), phosphinothricin (bar, isolated from Streptomyces hygroscopicus, codes for phophinothricin acetyltransferase), etc.

Step: Transforming Target Plants with the Gene Construct

There are two ways of genetically transforming the target plant:

1. Vector mediated gene transfer,
2. Vector less or direct gene transfer.

Step: Selection of The Transgenic Plant Tissue/Cells

Following the gene insertion process, plant tissues are transferred to a selective medium containing an antibiotic or herbicide, depending on which selectable marker was used. Only plants expressing the selectable marker gene will survive and it is assumed that these plants will also possess the transgene of interest.

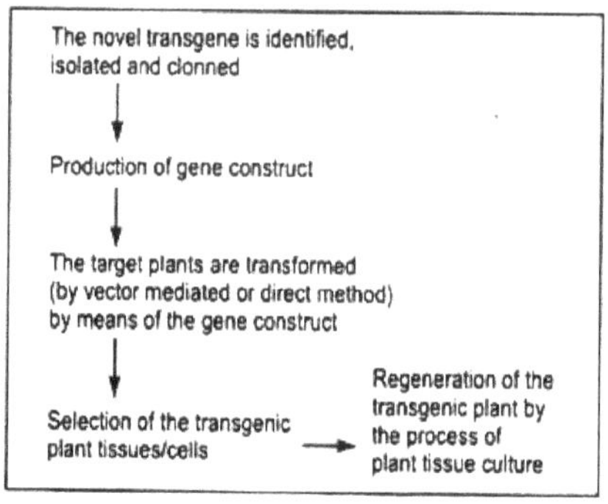

Step: Regeneration of the Transgenic Plants

To obtain whole plants from transgenic tissues, they are grown under controlled environmental conditions in a series of media containing nutrients and hormones by the process of plant tissue culture.

Promoters used in construction of gene construct

Promoter	Source	Relative Activity
35S	CaMV 35S RNA gene	Constitutive, high activity; most commonly used in dicots
35S + Adh1-I 1	35S promoter + first intron of maize Adh1 gene	Enhanced promoter activity; constitutive
35S + sh1-I 1	35S promoter + first intron of maize shrunken 1 gene	Better than 35S + Adh1-I 1 in monocots; constitutive
Adh1	Promoter of alcohol dehydrogenase gene of maize	Moderate activity in cereals; anaerobic expression
Emu	Modified from Adh1 promoter and its first intron	Moderate activity in cereals; anaerobic expression
Act1 + Act-I 1	Rice actin gene + its first intron	Moderate activity; constitutive
Ubi1 + Ubi1-I 1 (or I 6)	Maize ubiquitin 1 gene promoter + its first (or sixth) intron	High activity in cereals; constitutive
Vicilin promoter	Pea vicilin storage protein gene	Seed-specific promoter

Integration of the Transgene in the Genome of the Target Plant

In general, transgenes integrate at random sites in any of the chromosomes of the genome of host cells. Usually, in a given cell, integration occurs at a single location. As a result, different cells may be expected to show integration of the transgene at different chromosomal locations.

The number of copies integrated per genome ranges from one to several hundred. In general, multiple copies are integrated when large amounts of DNA are used for transfection, while single copies are integrated with smaller amounts.

When multiple copies are integrated, they are mostly integrated at one site joined to each other head-to-tail, i.e., as a concatemer. However, in a small proportion of cases, the multiple copies are located at several sites in the same genome.

The mechanism of random integration is not known. The entire gene construct, including the vector DNA, becomes integrated. When two different gene constructs are mixed and used for transfection, they tend to be integrated together at the same site; this is known as co-transfection. The sequences flanking a gene on either side influence the expression of this gene.

Therefore, the same transgene integrated at different locations in the genome may show different levels of expression; this is known as position effect. Transgene integration frequently leads to various forms of rearrangements, e.g., duplication, deletion, etc., near the site of integration.

If these changes are large enough, the host gene located at the site of integration may become non-functional. A host gene would also become non-functional if the transgene becomes integrated within the coding region of this gene. When integration of a transgene leads to the loss of function of a host gene, it is called insertional mutagenesis; it often produces aberrant phenotypes.

Analysis of Transgene Integration

The integration of transgene into the genome is confirmed by Southern hybridization of genomic DNA extracted from the considered transgenic individuals. The DNA is digested with a suitable restriction enzyme prior to electrophoresis.

By choosing appropriate restriction enzymes for DNA digestion, not only the integration of transgene can be established beyond doubt, but information on the number of copies per cell, the orientations of tandemly arranged copies and the presence of single or multiple integration sites is also obtained from Southern hybridization. All the individuals that give positive result with Southern hybridization are regarded as confirmed transgenic.

Detection of mRNA Expression

The mRNAs produced by transgenes is most readily detected if they are with unique sequences, which have no counterparts among those produced by the host genome. A high purity RNA preparation is obtained from the appropriate tissue of transgenic individuals, and is subjected to RNA dot blot hybridization with a radioactive probe specific for the transgene.

Alternatively, the RNA preparation may be used for northern hybridization, which provides additional information on transcript size as well.

Inheritance of Transgenes

The transgenes which are stably integrated are inherited in a Mendelian fashion. They are usually dominant. Instability may occur due to point mutation, like methylation, or rearrangements of the T-DNA region. In addition, homologous recombination between copies of the transgene inserted in the same nucleus can also lead to instability of the gene.

Future Development of Transgenic Technology

New techniques for producing transgenic plants will improve the efficiency of the process and will help resolve some of the environmental and health concerns.

Among the expected changes are the following:

- More efficient transformation, that is, a higher percentage of plant cells will successfully incorporate the transgene.
- Better marker genes to replace the use of antibiotic resistance genes.
- Better control of gene expression through more specific promoters, so that the inserted gene will be active only when and where needed.
- Transfer of multi-gene DNA fragments to modify more complex traits.

Cisgenic Plants

Cisgenic plants are made using genes found within the same species or a closely related one, where conventional plant breeding can occur. Some breeders and scientists argue that cisgenic modification is useful for plants that are difficult to crossbreed by conventional means (such as potatoes), and that plants in the cisgenic category should not require the same regulatory scrutiny as transgenics.

Cisgenesis is the production of genetically modified crops using a donor DNA fragment from the species itself or from a cross compatible species. The newly introduced gene is unchanged and includes its own introns and regulatory sequences and is free of any vector DNA, except T- DNA border sequences that flank the cisgene. The resultant phenotype of the cisgenic plant can be achieved through conventional breeding also; but it will take a much longer time. One of the most important points of cisgenesis is that it introduces only the desired gene, thus avoiding linkage drag that can be resulted from conventional cross breeding and also it eliminates hectic and time consuming backcrossing to recover the recurrent parent genotype. The concept of cisgenesis was introduced by Dutch researchers; Schouten, Krens and Jacobsen in the year 2006.

The worthiness of the GM techniques for developing highly reliable and good quality food products to the world has been set off by public worries about the safety of the derived food and there subsequent finished products. Most particularly, the controversy has spotlighted on the probable unpredictable hazards arising from the agglomeration of certain new substances in crop plants that confers toxicity, allergy and genetic threats to humans.

Benefits of Cisgenesis

- Cisgenesis is as safe as conventional breeding.
- To overcome the problem of linkage drag.
- Original genetic makeup of a crop is preserved any one or few genes are added.
- Escape of foreign genes via pollen grains to natural vegetation is not a problem in cisgenesis.
- In cisgenesis, only the desired genes are introduced without the undesirable genes unlike conventional breeding.

Drawbacks of Cisgenesis

- The gene outside the sexually compatible gene pool cannot be introduced.
- There is also a chance that the introduction of cisgene may influence the expression of genes that are already present in the recipient genome.
- Position effect may lead to alteration of the gene expression and phenotypic differences.
- The production of marker-free plants often requires the implementation or development of new techniques and such techniques may not be readily available.

Advantages of Cisgenesis over Conventional Breeding

1. Removes linkage drag: Introgression of new traits into the cultivated varieties by conventional methods comprises of wide crosses and repeated backcrossing. However, these traits are constantly linked with a large share of unwanted chromosomal region, the so called linkage drag. Some of these genes affect the normal features of the crop as they may engage in the production of diverse kinds of toxins or allergens. Hence, direct transfer of desired genes through cisgenesis into an existing variety without altering any of the properties enviable for the consumers can be accomplished.

2. Maintains the original genetic make-up of the plant varieties: In a hybridization method, the genetic makeup of the progeny plant varies from its parents because it has been a mixture of both the parental genomes. In spite of this, there is a necessity to conserve some part of the genome which reveals certain constructive traits. Through conventional plant breeding such an approach is not possible entirely due to self incompatibility among the vegetatively propagated plants. Cisgenic breeding tools are used in order to get up to four different resistance genes into one variety without changing the other traits of the modified variety.

3. Reduction in pesticide application: The key purpose of cisgenesis is to transfer disease resistance genes to susceptible varieties. The vital goal here is to lessen substantial pesticide application. As a result, there is decline in the input costs of the farmers and decreased pesticide leftovers on the plants and also in their products, which is mostly favoured by the consumers. This reduces the environmental pollution by pesticides and in turn helpes in sustainable agricultural development.

4. Time Saving: In conventional hybridization programmes, there is linkage drag, where there is inheritance of unwanted genes to the progeny. Several backcrossed generations are required to get rid of such kind of undesired genes. Cisgenesis overcomes the problem of linkage drag and only the gene of interest is introduced into the genome of the recipient plant within a short period of time. Thus, this saves a lot of time.

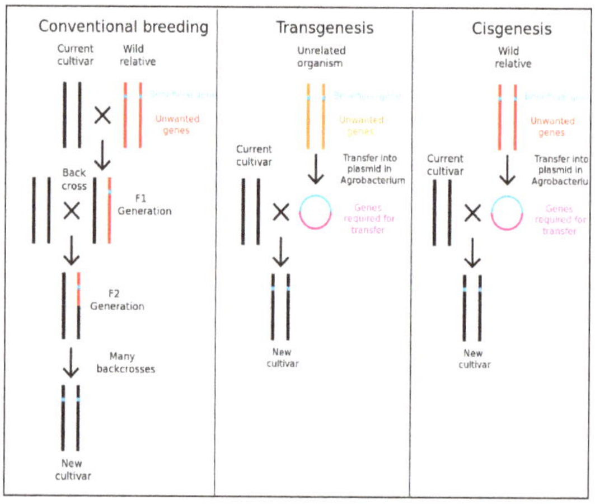

Subgenic

Genetically modified plants can also be developed using gene knockdown or gene knockout to alter the genetic makeup of a plant without incorporating genes from other plants. In 2014, Chinese researcher Gao Caixia filed patents on the creation of a strain of wheat that is resistant to powdery mildew. The strain lacks genes that encode proteins that repress defenses against the mildew. The researchers deleted all three copies of the genes from wheat's hexaploid genome. Gao used the TALENs and CRISPR gene editing tools without adding or changing any other genes. No field trials were immediately planned. The CRISPR technique has also been used to modify white button mushrooms (Agaricus bisporus).

Multiple Trait Integration

From a breeding standpoint, multiple trait integration (MTI) is a four-step process of converting an elite variety/hybrid for value-added traits (e.g. transgenic events) using backcross breeding, ultimately regaining the performance attributes of the target hybrid along with reliable expression of the value-added traits.

Protoplast Isolation

Each plant cell has a definite cellulose wall which is attached with each other by middle lamella made up of pectin. Protoplast is the content which is surrounded only by plasma-lemma. So to get the isolated protoplast it is essential to remove the pectin material to obtain the single cells and then to remove cell wall by enzymatic method.

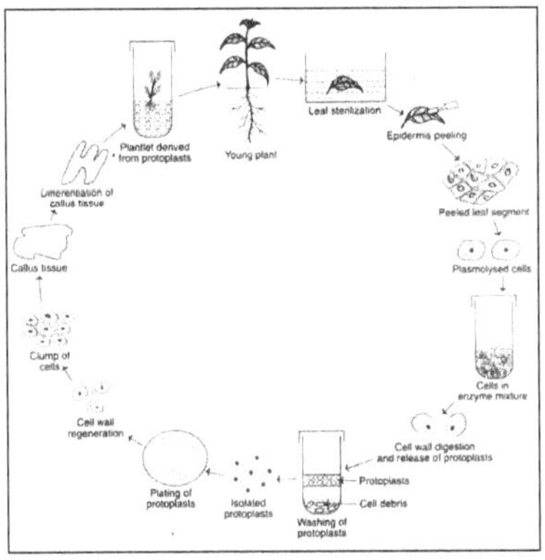

Sources of Explant for Protoplast Isolation

Protoplasts can be isolated directly from the different parts of whole plant which bears the soft parenchymatous tissue (e.g., young fully expanded soft leaves) or indirectly from the in vitro grown plant tissue (e.g., callus tissue). Protoplast yield and viability depends greatly on the type of tissue material from which it is isolated and the method used. Cell suspension cultures may also provide a very good source of protoplast.

Before isolation of protoplast, the source material if it is from in vivo grown plant then it should be properly surface sterilized using the proper method of sterilization. Then any of the methods either mechanical or enzymatic can be used to isolate the protoplast.

Methods of Protoplast Isolation

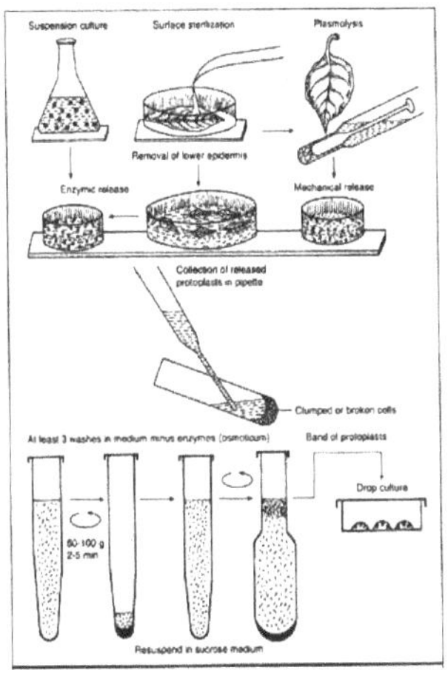

The essential step of protoplast isolation is the proper use of osmoticum. The plant protoplast is an osmotic system where the plasma membrane acts as the semipermeable membrane which equilibrates the outward and inward flow of water. If the inward pressure is more than due to heavy inflow of water the protoplast will burst.

So during protoplast isolation the enzymes applied to isolate the protoplast and the tissue material should be placed in proper osmolyticum or plasmolyticum. Generally mannitol, a sugar alcohol at high concentration, an easily transportable component provides a good stable osmotic environment which prevents bursting of protoplast.

Protoplasts may be isolated by any one of the two following ways:

- Mechanical method (Non-enzymatic method);
- Enzymatic method.

Mechanical Method

Any soft parenchymatous tissue is kept in a plasmolyticum. The plasmolysed tissue is then finely chopped into pieces and the intact cells (plasmolysed) are released into the medium from the cut surface. The suspension is then allowed for deplasmolysis and the released protoplasts attain their original size. The yield of protoplast in this method is very low, for large scale of protoplast yield the enzymatic method is followed.

Enzymatic Method

Young fully expanded soft leaves, or in vitro grown callus tissue or cell suspension culture grown cells can be used as the source material. The tissues or cells are incubated in plasmolyticum for 1 hr before enzymatic treatment. The intact tissue materials cut into smaller pieces to increase the surface area of enzymatic activity. The enzymes can be used either sequentially in two step method or in a single step by mixed enzymatic method.

The enzymes used are of three main categories:

- Cellulases:
  - Cellulysin
  - Driselase
- Hemicellulase:
  - Helicase
  - Rhozyme
- Pectinase:
  - Macerozyme-R-10
  - Pectinol

The concentration of enzymes used and the time period of incubation varies greatly depending on the tissue type. In two step method, the pectinase and hemicellulases are applied first and then cellulase is applied for complete removal of cell wall.

After release of protoplast into the suspension, for removal of enzymes the protoplasts are collected in centrifuge tube as pellet and washed several times with the osmoticum.

Purification of Protoplasts

For purification, the protoplasts suspended in osmoticum are centrifuged using sucrose (20%) solution. The viable protoplasts float on the top surface of sucrose solution forming a band. These protoplasts are then collected, re-suspended in osmoticum and washed several times.

Finally the protoplasts suspended in a measured volume of protoplast culture medium after counting the number with the help of hemocytometer. The viability of protoplast is checked with the help of Fluorescin diacetate staining or phenosafranine or calcofluore white.

Isolation of Sub-Protoplasts

Sub-protoplasts do not contain the entire content of plant cells, these can be of different types such as cytoplasts, karyoplasts or mini-protoplasts or micro-protoplasts or microplasts.

- Cytoplasts lack the nucleus and contain the entire cytoplasm of a cell or part of it. These are often used for cybrid production.
- Karyoplasts or Mini-protoplasts contain a nucleus surrounded by some cytoplasm and the original outer plasma membrane.
- Micro-protoplasts contain fraction of both nuclear and cytoplasmic material.
- Microplasts lack the nuclear material and contain only fraction of cytoplasm and outer membrane.

Protoplast Viability Test

The most frequently used staining methods for checking protoplast viability are:

1. Fluorescein diacetate (FDA) dissolved in acetone is used at a conc. of 0.01% and intact viable protoplasts only fluoresce when observed under UV.
2. Phenosafranine is also used at a conc. of 0.01 %, which is specific for dead protoplast that shows red in colour.

Somatic Cell Hybridization

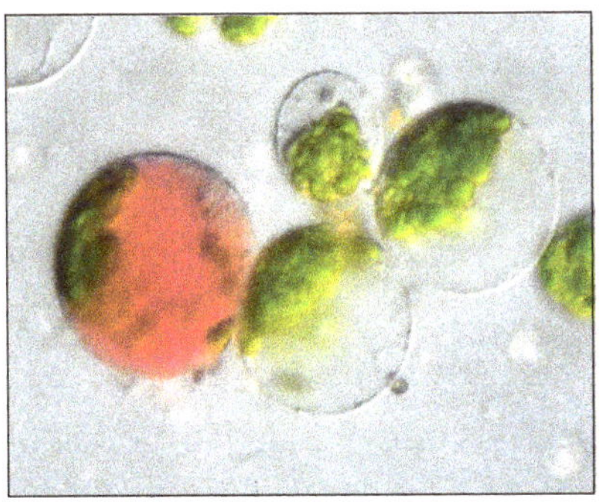

The process by which protoplasts of two different plant species fuse together to form hybrids is known as somatic hybridisation and the hybrids so produced is known as somatic hybrids.

Sexual hybridization since time immemorial has been used as a method for crop improvement but it has its own limitations as it can only be used within members of same species or closely related wild species. Thus, this limits the use of sexual hybridization as a means of producing better varieties. Development of viable cell hybrids by somatic hybridization, therefore, has been considered as an alternative approach for the production of superior hybrids overcoming the species barrier. The technique can facilitate breeding and gene transfer, bypassing problems associated with conventional sexual crossing such as, interspecific, intergeneric incompatibility. This technique of hybrid production via protoplast fusion allows combining somatic cells (whole or partial) from different cultivars, species or genera resulting in novel genetic combinations including symmetric somatic hybrids, asymmetric somatic hybrids or somatic cybrids.

The most common target using somatic hybridization is the gene of symmetric hybrids that contain the complete nuclear genomes along with cytoplasmic organelles of both parents. This is unlike sexual reproduction in which organelle genomes are generally contributed by the maternal parent. On the other hand, somatic cybridization is the process of combining the nuclear genome of one parent with the mitochondrial and/or chloroplast genome of a second parent. Cybrids can be produced by donor-recipient method or by cytoplast-protoplast fusion. Incomplete asymmetric somatic hybridization also provides opportunities for transfer of fragments of the nuclear genome, including one or more intact chromosomes from one parent (donor) into the intact genome of a second parent (recipient).

These methods provide genetic manipulation of plants overcoming hurdle of sexual incompatibility, thereby, serving as a method of bringing together beneficial traits from taxonomically distinct species which cannot be achieved by sexual crosses. Several parameters such as, source tissue, culture medium and environmental factors influence ability of a protoplast derived hybrid cells to develop into a fertile plant. The general steps involved in somatic hybridization and cybridization methods are elaborated in figures below.

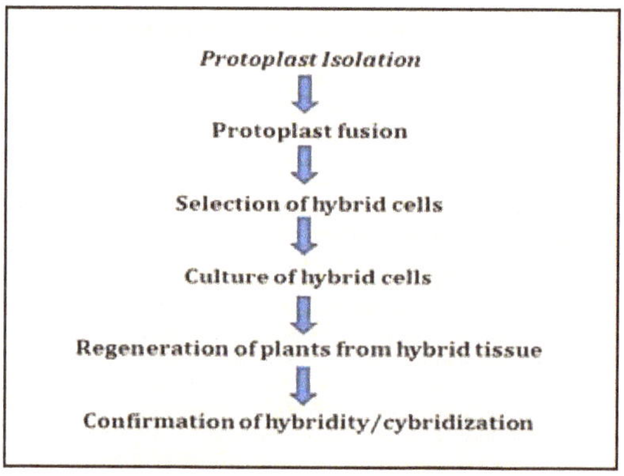

Steps Involved in Somatic Hybridization.

Protoplast Fusion

Protoplast fusion could be spontaneous during isolation of protoplast or it can be induced by mechanical, chemical and physical means. During spontaneous process, the adjacent protoplasts fuse together as a result of enzymatic degradation of cell walls forming homokaryons or homokaryocytes, each with two to several nuclei. The occurrence of multinucleate fusion bodies is more frequent when the protoplasts are prepared from actively dividing callus cells or suspension cultures. Since the somatic hybridization or cybridization require fusion of protoplasts of different origin, the spontaneous fusion has no value. To achieve induced fusion, a suitable chemical agent (fusogen) like, NaNO3, high Ca2+, polyethylene glycol (PEG), or electric stimulus is needed.

i. Fusion by means of NaNO3: It was first demonstrated by Kuster in 1909 that the hypotonic solution of NaNO3 induces fusion of isolated protoplast forming heterokaryon (hybrid). This method was fully described by Evans and Cocking, however this method has a limitation of generating few no of hybrids, especially when highly vacuolated mesophyll protoplasts are involved.

ii. High pH and Ca++ treatment: This technique lead to the development of intra- and interspecific hybrids. It was demonstrated by Keller and Melcher in 1973. The isolated protoplasts from two plant species are incubated in 0.4 M mannitol solution containing high Ca++(50 mM CaCl2.2H2O) with highly alkaline pH of 10.5 at 37°C for about 30 min. Aggregation of protoplasts takes place at once and fusion occurs within 10 min.

iii. Polyethylene glycol treatment: Polyethylene glycol (PEG) is the most popularly known fusogen due to ability of forming high frequency, binucleate heterokaryons with low cytotoxicity. With PEG the aggregation occurred mostly between two to three protoplasts unlike Ca++ induced fusion which involves large clump formation. The freshly isolated protoplasts from two selected parents are mixed in appropriate proportions and treated with 15-45% PEG (1500-6000MW) solution for 15-30 min followed by gradual washing of the protoplasts to remove PEG. Protoplast fusion occurs during washing. The washing medium may be alkaline (pH 9-10) and contain a high Ca++ ion concentration (50 mM). This combined approach of PEG and Ca++ is much more efficient than the either of the treatment alone. PEG is negatively charged and may bind to cation like Ca++, which in turn, may bind to the negatively charged molecules present in plasma lemma, they can also bind to cationic molecules of plasma membrane. During the washing process, PEG molecules may pull out the plasma lemma components bound to them. This would disturb plamalemma organization and may lead to the fusion of protoplasts located close to each other. The technique is nonselective thus, induce fusion between any two or more protoplasts.

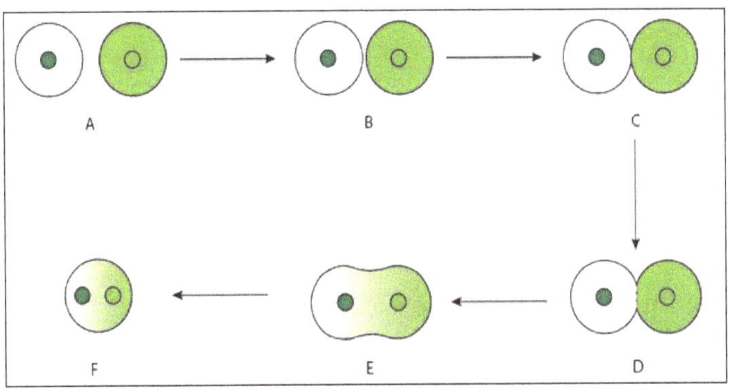

Figure: Sequential stages in protoplast fusion. (A) two separate protoplasts, (B) agglutination of two protoplasts, (C and D) Membrane fusion at localized site, and (E and F) development of spherical heterokaryon.

iv. Electrofusion: The chemical fusion of plant protoplast has many disadvantages – (1) The fusogen are toxic to some cell systems, (2) it produces random, multiple cell aggregates, and (3) must be removed before culture. Compare to this, electrofusion is rapid, simple, synchronous and more easily controlled. Moreover, the somatic hybrids produced by this method show much higher fertility than those produced by PEG-induced fusion.

Zimmermann and Scheurich demonstrated that batches of protoplasts could be fused by electric fields by devising a protocol which is now widely used. This protocol involves a two-step process. First, the protoplasts are introduced into a small fusion chamber containing parallel wires or plates which serve as electrodes. Second, a low-voltage and rapidly oscillating AC field is applied, which causes protoplasts to become aligned into chains of cells between electrodes. This creates complete cell-to-cell contact within a few minutes. Once alignment is complete, the fusion is induced by application of a brief spell of high-voltage DC pulses (0.125-1 kVcm-1). A high voltage DC pulses induces a reversible breakdown of the plasma membrane at the site of cell contact, leading to fusion and consequent membrane reorganization. The entire process can be completed within 15 min.

Selection of Fusion Products

The somatic hybridization by electrofusion of protoplasts allow one-to-one fusion of desired pairs of protoplasts and, therefore, it is easy to know the fate of fusion products. However, protoplast suspension recovered after chemical treatments (fusogen) consists of the following cell types:

- unfused protoplasts of the two species/strains,
- products of fusion between two or more protoplasts of the same species (homokaryons),
- 'hybrid' protoplasts produced by fusion between one (or more) protoplasts of each of the two species (heterokaryons).

The heterokaryons which are the potential source of future hybrids constitute of a very small (0.5-10%) proportion of the mixture. Therefore, an effective strategy has to be employed for their identification and isolation. Various protocols have been proposed and practiced for the effective selection of hybrids, including morphological basis, complementation of biochemical and genetic traits of the fusing partners, and manual or electronic sorting of heterokaryons/hybrid cells.

Morpho-physiological basis: The whole mixtures of the protoplasts are cultured after fusion treatment and the resulting calli or regenerants are screened for their hybrid characteristics. Occasionally the hybrid calli outgrow the parental cell colonies and are identified by their intermediate morphology, i.e. green with purple coloured cells. However, the process is labour intensive and requires glasshouse facilities. It is limited to certain combinations showing differences in their regeneration potential under specific culture conditions.

Complementation: In this case complementation or genetic or metabolic deficiencies of the two fusion partners are utilized to select the hybrid component. When protoplasts of two parents, (one parent bearing cytoplasmic albino trait and the other parent bearing green trait) each parent carrying a non-allelic genetic or metabolic defect are fused, it reconstitutes a viable hybrid cell, of wild type in which both defects are mutually abolished by complementation, and the hybrid cells are able to grow on minimal medium non-permissive to the growth of the parental cells bearing green trait. Later, the calli of hybrid nature could be easily distinguished from the parental type tissue (albino trait) by their green color. The complementation selection can also be applied to dominant characters, such as dominant resistance to antibiotics, herbicides or amino acid analogues.

Isolation of heterokaryons or hybrid cells: The manual or electronic isolation of heterokaryons or hybrid cells is the most reliable method. Manual isolation requires that the two parental type protoplasts have distinct morphological markers and are easily distinguishable. For example, green vacuolated, mesophyll protoplasts from one parent and richly cytoplasmic, non green protoplasts from cultured cells of another parent. The dual fluorescence method also helps easy identification of fusion products. In this case, the protoplast labeled green by treatment with fluorescein diacetate (FDA, 1-20 mgl-1) are fused with protoplasts emitting a red fluorescence, either from chlorophyll autofluorescence or from exogenously applied rhodamine isothiocyanate (10-20 mgl-1). The labeling can be achieved by adding the compound into the enzyme mixture. This can be applied even for morphologically indistinguishable protoplasts from two parents.

Verification and Characterization of Somatic Hybrids

As no system is foolproof and they have their own advantages and disadvantages. Therefore, even after selecting the desired hybrids/cybrids following protoplast fusion, it is required to carry out one or more tests to compare the parent protoplast lines with the putative hybrids. Some of the techniques that can be tried are:

Morphology: Somatic hybrids in most of the cases show characters intermediate between the two parents such as, shape of leaves, pigmentation of corolla, plant height, root morphology and other vegetative and floral characters. The method is not much accurate as tissue culture conditions may also alter some morphological characters or the hybrid may show entirely new traits not shown by any of the parents.

Isozyme analysis: Multiple molecular forms of same enzyme which catalyses similar or identical reactions are known as isozymes. Electrophoresis is performed to study banding pattern as a check for hybridity. If the two parents exhibit different band patterns for a specific isozyme the putative hybrid can be easily verified. The isozymes commonly used for hybrid identification include, acid phosphatase, esterase, peroxidase.

Cytological analysis: Chromosome counting of the hybrid is an easier and reliable method to ensure hybridity as it also provides the information of ploidy level. Cytologically the chromosome count of the hybrid should be sum of number of chromosomes from both the parents. Besides number of chromosomes, the size and structure of chromosomes can also be monitored. However, the approach is not applicable to all species, particularly where fusion involves closely related species or where the chromosomes are very small. Moreover, sometimes the somaclonal variations may also give rise to different chromosome number.

Molecular analysis: Specific restriction pattern of nuclear, mitochondrial and chloroplast DNA characterizes the plastomes of hybrids and cybrids. Molecular markers such as RFLP, RAPD, ISSR can be employed to detect variation and similarity in banding pattern of fused protoplasts to verify hybrid and cybrid.

Cybrids or Cytoplasmic Hybrids

Sexual hybridization involves fusion of the nuclear genes of both the parents but somatic hybrids involves even cytoplasm from both the parental species in hybrid obtained by protoplast fusion. However, in another case somatic hybrids containing nuclear genome of one parent but cytoplasm from both the parents, are termed as cybrids. The approach is time consuming and require several years of crossing plants provides an opportunity to study interparental mitochondrial, chloroplast fusion giving rise to plants with novel genomes.

Methods to Produce Cybrids

They are produced in variable frequencies in normal protoplast fusion experiments due to one of the following methods:

1. Fusion of normal protoplast with an enucleated protoplast. The enucleated protoplast can be produced by high speed centrifugation (20,000-40,000xg) for 60 min with 5-50% percoll.
2. Fusion between a normal protoplast and another protoplast with a non-viable nucleus or suppressed nucleus.
3. Elimination of one of the nuclei after heterokaryons formation.
4. Selective elimination of chromosomes at a later stage.
5. Irradiating (with X-rays or gamma rays) the protoplasts of one species prior to fusion in order to inactivate their nuclei.
6. By preparing enucleate protoplasts (cytoplasts) of one species and fusing them with normal protoplasts of the other species.

Cybrids provide the following unique opportunities: (i) transfer of plasmogenes of one species into the nuclear background of another species in a single generation, and even in (ii) sexually incompatible combinations, (iii) recovery of recombinants between the parental mitochondrial or chloroplast DNAs (genomes), and (iv) production of a wide variety of combinations of the parental and recombinant chloroplasts with the parental or recombinant mitochondria.

Applications of Somatic Hybridization

1. Novel interspecific and intergeneric crosses which are difficult to produce by conventional methods can be easily obtained.
2. Important characters, such as resistance to diseases, ability to undergo abiotic stress and other quality characters, can be obtained in hybrid plant by the fusion of protoplasts of plant bearing particular character to the other plant which may be susceptible to diseases.
3. Protoplasts of sexually sterile haploid, triploid, aneuploid plants can be fused to obtain fertile diploids and polyploids.
4. Studying cytoplasmic genes may be helpful to carry out plant breeding.

5. Most of the agronomically important traits, such as cytoplasmic male sterility, antibiotic resistance and herbicide resistance, are cytoplasmically encoded, hence can be easily transferred to other plant.

6. Plants in juvenile stage can also be hybridized by means of somatic hybridization.

7. Somatic hybridization can be used as a method for the production of autotetraploids.

Limitations of Somatic Hybridization

1. Application of protoplast methodology requires efficient plant regeneration system from isolated protoplasts. Protoplasts from two species can be fused, however, production of somatic hybrids is not easy.

2. Lack of a proper selection method for fused products (hybrids) poses a problem.

3. The end product of somatic hybridization are often unbalanced (sterile, misformed and unstable).

4. Somatic hybridization of two diploids leads to formation of amphidiploids which is unfavorable.

5. It is not sure for a character to completely express after somatic hybridization.

6. The regeneration products of somatic hybridization are often variable due to somaclonal variation, chromosome elimination, organelle segregation.

7. All diverse intergeneric somatic hybrids are sterile and, therefore, have limited chances of development of new varieties.

8. To transfer useful genes from wild species to cultivated crop, it is necessary to achieve intergeneric recombination or chromosome substitution between parental genomes.

Gene Gun Method

The gene gun method is a method used for genetically modifying plants.

With the use of a gene gun, the gene gun method delivers extra DNA directly into a plant's nucleus. The method is also commonly called particle acceleration or microprojectile bombardment.

The gene gun can be used on seedlings or tissue culture cells. Prior to injecting the DNA into the plant tissue via the gene gun, either microscopic gold or tungsten particles are liberally coated with many hundreds of copies of genes. The particles are then forced into the nucleus with the gene gun.

The first gene gun method of DNA modification was developed in 1980. The process relied on a 22-caliber cartridge to shoot the gene-coated metal particles into the plant's cell.

Nowadays, modern versions of the gene gun operate differently. The modernized gene gun depends on a vacuum chamber and high-pressure gas to hurl the metal particles into the plant cells.

With each firing of the gene gun, the pressurized gas is released in a sudden popping burst and the gene coated metal particles fire into the cells.

The gene gun is highly effective at modifying the DNA and genetics of plant cells.

Genetic modification in plants can make them resistant to drought, disease, and pests. It also helps create a more nutritious edible plant with higher levels of proteins.

This method was used to create several commercial events such as RoundupReady soybean, it has several limitations that make it less appealling than the use of Agrobacterium as a transformation method.

The gene gun is a good example of a creative idea being developed into a practical technology.

The gene gun can be used on either tissue culture cells or seedlings (to make chimeric plants) of any species. As the name implies, this method works by shooting DNA into the plant cells.

Gene Gun Design

The gene gun was originally a Crosman air pistol modified to fire dense tungsten particles. It was invented by John C Sanford, Ed Wolf and Nelson Allen at Cornell University, and Ted Klein of DuPont, between 1983 and 1986. The original target was onions (chosen for their large cell size) and it was used to deliver particles coated with a marker gene. Genetic transformation was then proven when the onion tissue expressed the gene.

The earliest custom manufactured gene guns (fabricated by Nelson Allen) used a 22 caliber nail gun cartridge to propel an extruded polyethylene cylinder (bullet) down a 22 cal. Douglas barrel. A droplet of the tungsten powder and genetic material was placed on the bullet and shot down the barrel at a lexan"stopping" disk with a petri dish below. The bullet welded to the disk and the genetic material blasted into the sample in the dish with a doughnut effect (devastation in the middle, a ring of good transformation and little around the edge). The gun was connected to a vacuum pump and was under vacuum while firing. The early design was put into limited production by a Rumsey-Loomis (a local machine shop then at Mecklenburg Rd in Ithaca, NY, USA). Later the design was refined by removing the "surge tank" and changing to nonexplosive propellants. DuPont added a plastic extrusion to the exterior to visually improve the machine for mass production to the scientific community. Bioradcontracted with Dupont to manufacture and distribute the device. Improvements include the use of helium propellant and a multi-disk-collision delivery mechanism. Other heavy metals such as gold and silver are also used. Gold may be favored because it has better uniformity than tungsten and tungsten can be toxic to cells, but its use may be limited due to availability and cost.

Biolistic Construct Design

A construct is a piece of DNA inserted into the target's genome, including parts that are intended to be removed later. All biolistic transformations require a construct to proceed and while there is great variation among biolistic constructs, they can be broadly sorted into two categories: those which are designed to transform eukaryotic nuclei, and those designed to transform prokaryotic-type genomes such as mitochondria, plasmids or plastids.

Those meant to transform prokaryotic genomes generally have the gene or genes of interest, at least one promoter and terminator sequence, and a reporter gene; which is a gene used to enable detection or removal of those cells which didn't integrate the construct into their DNA. These genes may each have their own promoter and terminator, or be grouped to produce multiple gene products from one transcript, in which case binding sites for translational machinery should be placed between each to ensure maximum translational efficiency. In any case the entire construct is flanked by regions called border sequences which are similar in sequence to locations within the genome, this allows the construct to target itself to a specific point in the existing genome.

Constructs meant for integration into a eukaryotic nucleus follow a similar pattern except that: the construct contains no border sequences because the sequence rearrangement that prokaryotic constructs rely on rarely occurs in eukaryotes; and each gene contained within the construct must be expressed by its own copy of a promoter and terminator sequence.

Though the above designs are generally followed, there are exceptions. For example, the construct might include a Cre-Lox system to selectively remove inserted genes; or a prokaryotic construct may insert itself downstream of a promoter, allowing the inserted genes to be governed by a promoter already in place and eliminating the need for one to be included in the construct.

Application

Gene guns are so far mostly used with plant cells. However, there is much potential use in humans and other animals as well.

Plants

The target of a gene gun is often a callus of undifferentiated plant cells or a group of immature embryos growing on gel medium in a Petri dish. After the DNA-coated gold particles have been delivered to the cells, the DNA is used as a template for transcription (transient expression) and sometimes it integrates into a plant chromosome ('stable' transformation).

If the delivered DNA construct contains a selectable marker, then stably transformed cells can be selected and cultured using tissue culture methods. For example, if the delivered DNA construct contains a gene that confers resistance to an antibiotic or herbicide, then stably transformed cells may be selected by including that antibiotic or herbicide in the tissue culture media.

Transformed cells can be treated with a series of plant hormones, such as auxins and gibberellins, and each may divide and differentiate into the organized, specialized, tissue cells of an entire plant. This capability of total re-generation is called totipotency. The new plant that originated from a successfully transformed cell may have new traits that are heritable. The use of the gene gun may be contrasted with the use of Agrobacterium tumefaciens and its Ti plasmid to insert DNA into plant cells.

Advantages

Biolistics has proven to be a versatile method of genetic modification and it is generally preferred to engineer transformation-resistant crops, such as cereals. Notably, Bt maize is a product of biolistics. Plastid transformation has also seen great success with particle bombardment when compared to other current techniques, such as Agrobacterium mediated transformation, which have difficulty targeting the vector to and stably expressing in the chloroplast. In addition, there are no reports of a chloroplast silencing a transgene inserted with a gene gun. Additionally, with only one firing of a gene gun, a skilled technician can generate two transformed organisms. This technology has even allowed for modification of specific tissues in situ, although this is likely to damage large numbers of cells and transform only some, rather than all, cells of the tissue.

Limitations

Biolistics introduces DNA randomly into the target cells. Thus the DNA may be transformed into whatever genomes are present in the cell, be they nuclear, mitochondrial, plasmid or any others, in any combination, though proper construct design may mitigate this. Another issue is that the gene inserted may be overexpressed when the construct is inserted multiple times in either the same or different locations of the genome. This is due to the ability of the constructs to give and take genetic material from other constructs, causing some to carry no transgene and others to carry multiple copies; the number of copies inserted depends on both how many copies of the transgene an inserted construct has, and how many were inserted. Also, because eukaryotic constructs rely on illegitimate recombination, a process by which the transgene is integrated into the genome without similar genetic sequences, and not homologous recombination, which inserts at similar sequences, they cannot be targeted to specific locations within the genome, unless the transgene is co-delivered with genome editing reagents.

Electroporation

Electroporation is a popular physical method for introducing new genes directly into the protoplasts.

In this method, electric field is playing important role. Due to the electric field protoplast get temporarily permeable to DNA. In electroporation, plant cell protoplasts are kept in an ionic solution containing the vector DNA in a small chamber that has electrodes at opposite ends. A pulse of high voltage is applied to the electrode which makes the transient pores (ca. 30 nm) in the plasma membrane, allowing the DNA to diffuse into the cell. Immediately, the membrane reseals.

If appropriately treated, the cells can regenerate cell wall, divide to form callus and, finally, regenerate complete plants in suitable medium. The critical part of the procedure is to determine conditions which produce pores that are sufficiently large and remain open long enough to allow for DNA diffusion. At the same time, the conditions should make pores that are temporary. With a 1 cm gap between the electrodes and protoplasts of 40-44μm diameter, 1-1.5 kVcm -2 of field strength for 10μs is required for efficient introduction of DNA. It was seen that presence of 13% PEG (added after DNA) during electroporation significantly raised the transformation frequency. The other factors which may improve the transformation frequency by electroporation are linearizing of plasmid, use of carrier DNA, and heat shock (45 ~ for 5 min) prior to addition of vector, and placing on ice after pulsing. Under optimal conditions transformation frequencies of up to 2% have been reported. Stably transformed cell lines and full plants of a number of cereals have been produced through electroporation.

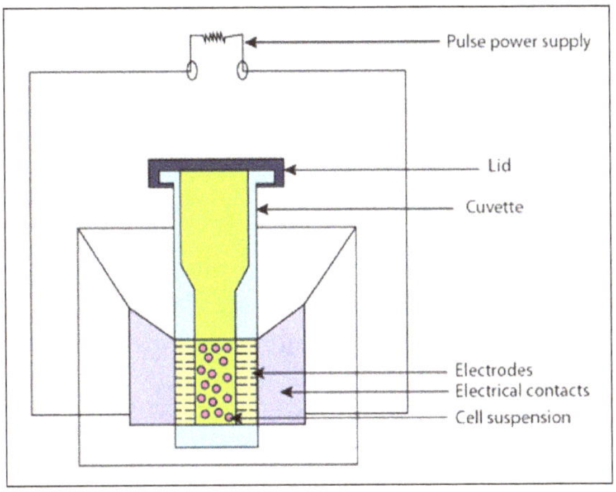

Figure: Electroporation.

There are some parameters that can be considered when performing in vitro electroporation:

Cell Size

Cell size is inversely correlated to the size of the external field needed to generate permeabilization. Consequently, optimization for each cell type is essential. Likewise, cell orientation matters for cells that are not spherical.

Temperature

It has been observed that plant membrane resealing is effectively temperature dependent and shows slow closure at low temperatures. For DNA transfer, it has been found that cooling at the time of permeabilization and subsequent heating in incubator increases transfer efficacy and cell viability.

Post-pulse Manipulation

Cells are susceptible when in the permeabilized state, and it has been shown that waiting for 15min after electroporation in order to allow resealing before pipetting cells, increases cell viability.

Composition of Electrodes and Pulsing Medium

For short pulses is needed for release of metal from the standard aluminium electrodes used in standard disposable cuvettes. Some authors advocate the use of low conductivity or more resistance media for DNA transfer in order to increase viability and increase transfection efficacy.

Microinjection

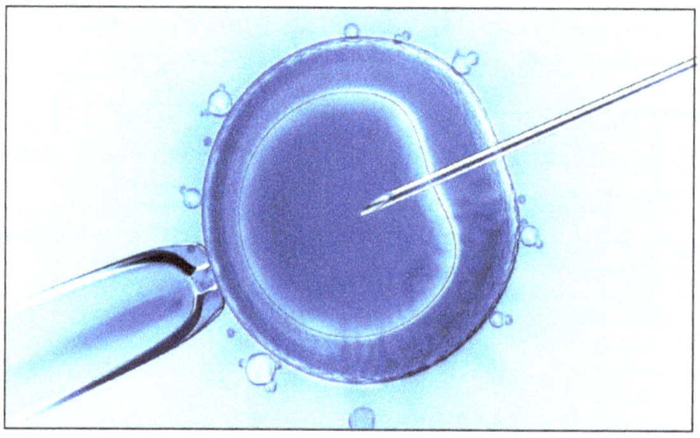

Microinjection is an established method to introduce DNA into single cells to generate transient and stable transformants. The injection of single cells within plant organs, however, is more difficult, and only a low percentage of the injected cells survive. The Eppendorf micromanipulation system has been used to inject tomato, mustard and Arabidopsis seedlings.

Usually plant cells possess a very rigid cell wall, which is a problem for microinjection since it requires stable injection capillaries with small diameters. In combination with the cellular turgor, the wall is responsible for the structural integrity of a plant seedling. The cell turgor of plants can be very high and is not only responsible for the morphology of a seedling but also plays an important role in development. Even if the injection against the high internal pressure was successful, the injury of a larger cell wall area will automatically lead to the destruction of the cell, at the latest when the needle is retreated.

The large central vacuole of a plant cell is also a problem that diminishes the chances of a successful cytoplasmic injection. The cytoplasm surrounding the central vacuole is only a few micrometres thick. As a result of the deformation of the cell wall during the injection and the sudden relaxation when the capillary penetrates the wall, the needle will often end up in the vacuole. Material injected into the vacuole is not only trapped but in many cases degraded, due to the low pH and the activity of several proteases. In addition, the tonoplast – the single membrane surrounding the vacuole – is easily injured, which leads to leakage of cytotoxic substances into the cytoplasm and hence to cell death.

Furthermore, because of multiple cell layers, light scattering and pigmentation, the optical properties of whole plant seedlings or plant organs are not optimal. Therefore, it is often difficult to identify different cell layers or even subcellular structures like vacuoles under microscopic control during the microinjection process.

To approach the described problems, Eppendorf developed a micromanipulation system equipped with very fast micromotors, variable injection parameters, high holding and injection pressure, in combination with a piezo step function. The set-up was tested for its suitability to inject plant cells – specifically tomato, mustard and Arabidopsis seedlings.

The micromanipulation system's injection angle was adjusted to 35° and axial injection movements were performed at a speed of 3000 μm/s with a step size of 10–20 μm. These settings ensure minimal injuries of the cell wall and thus raise the survival rate after injection. The injection movement can be supported by synchronised piezo impulses when the system's InjectMan 4 and PiezoXpert devices are electronically connected. As a marker for injection, Lucifer yellow (ly) – a water-soluble dye that easily allows discrimination between cytoplasmic and vacuolar localised fluorescence – was injected.

Typically, 5–8 injections per seedling were performed into cells of the etiolated hypocotyl. In the case of the tomato and mustard seedling injections, efficiencies of 26% and 48% were obtained. The injection efficiency for Arabidopsis seedlings was about 10% at 24 h after injection – strongly decreased compared to the tomato and mustard seedlings. However, Arabidopsis seedlings are much smaller and very fragile. As a consequence, injected cells do not only die relatively often but in many cases the whole hypocotyl is affected. Considering these challenges, 10% injection efficiency reflects a good value for Arabidopsis seedlings in our opinion.

Localised fluorescence of Iy 24 h after injection indicated that the cells were viable, but ly is no in vivo marker. Hence, we decided to inject Alexa Fluor 488-labelled histone (histone-AF488). After injection into the cytoplasm, the histones should be transported into the nucleus, where they accumulate. Histone import into the nucleus is an active process that requires GTP and intact protein complexes. Therefore, the import of histone-AF488 should only occur in living cells. The percentage of cells showing histone-AF488 fluorescence was in the same range as the observed percentage after ly injection, confirming that these cells survived the injection procedure.

The Eppendorf micromanipulation system, with its fast and precise motor movements, was thus well suited for the injection of plant cells. This includes the injection of Arabidopsis seedlings with relatively high efficiencies. Since a large collection of Arabidopsis mutants and transgenic lines is available for many research areas, microinjection of Arabidopsis cells from specific organs and tissues should, in addition to injection of other plant species, help to analyse numerous scientific questions.

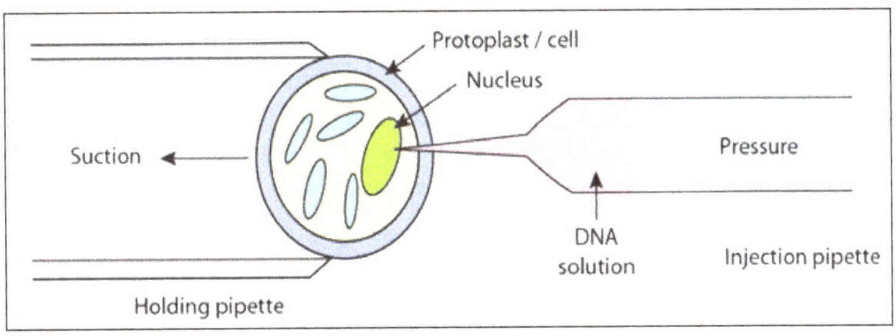

Agrobacterium

Agrobacterium elicits neoplastic growths (called crown gall tumors) that affect most dicotyledonous plants. Moreover, although plants represent the natural hosts for Agrobacterium, this microorganism can also genetically transform a wide range of other eukaryotic species, from yeast to mushrooms and filamentous fungi to phytopathogenic fungi to human cells. Most functions for Agrobacterium-host cell DNA transfer are coded by a large (200-kb) tumor-inducing (Ti) plasmid that resides in the bacterial cell and carries two important genetic components: the vir (virulence) region and the T-DNA delimited by two 25-bp direct repeats at its ends, termed the T-DNA borders. The vir region comprises seven major loci, virA, virB, virC, virD, virE, virG, and virH, which encode most of the bacterial protein machinery (Vir proteins) of the DNA transport. After induction of vir gene expression by small phenolic signal molecules secreted from wounded susceptible plant cells, the T-DNA borders are nicked by the bacterial VirD2 endonuclease, generating a transferable single-stranded (ss) copy of the bottom strand of the T-DNA region, designated the T strand.

Interestingly, the T strand does not travel alone but is thought to directly associate with two Agrobacterium proteins, VirD2 and VirE2, forming a transport (T) complex in which one molecule of VirD2 is covalently attached to the 5′-end of the T strand, whereas VirE2, an ssDNA-binding protein, is presumed to cooperatively coat the rest of the T strand molecule. Although only the wild-type T-DNA carries Ti genes, any DNA placed between the T-DNA borders will be transferred to the plant host. This lack of sequence specificity implies that a T-DNA molecule itself does not encode protein machinery for its transport from the bacterial cell into the host cell, import into the host cell nucleus, and integration into the host cell genome. Instead, these functions are fulfilled by the bacterial Vir proteins and their host cell partners.

In the last quarter of a century, since the discovery of the stable integration of the bacterial DNA in crown gall tumors, Agrobacterium has served as a primary a tool for plant genetic engineering. Furthermore, the Agrobacterium-host cell interaction also represents a unique and powerful experimental system to study a wide spectrum of basic biological processes such as cell-cell recognition and cell-to-cell transport, nuclear import, assembly and disassembly of protein-DNA complexes, DNA recombination, and regulation of gene expression.

Genetic Engineering

Predating the science fiction visions of Nano machines performing genetic engineering and other biotechnological tasks, Agrobacterium is a present-day microscopic but extremcly complex machine routinely used to alter genotypes of higher plants. This use of Agrobacterium is based on its unique capacity for "trans-kingdom sex", i.e. transfer of genetic material between prokaryotic and eukaryotic cells. Decades of research altered, augmented, and vastly improved this natural capacity, resulting in ingeniously modified Agrobacterium strains that can transfer and stably integrate virtually any gene to a variety of plant species, from research model plants such as Arabidopsis to agriculturally important rice and corn. In this issue, Valentine reviews the molecular mechanisms of T-DNA transfer that underlie the use of Agrobacterium as gene vector. Importantly, a special emphasis is made on the ethical and political aspects of the release of Agrobacterium-generated genetically modified plants into the environment and on the attitudes of different and often opposing geopolitical and economic forces to these matters.

Cell-Cell Recognition and Attachment

Agrobacterium recognition of and attachment to the host cells is an early and essential step of the infection process. The bacterial proteins participating in these events are encoded by several loci, e.g. chvA, chvB, pscA, and att. In particular, the att genes, such as attR and attD, are located on a cryptic plasmid pAtC58. Although pAtC58 is about twice as large as the Ti plasmid and, thus, likely encodes numerous bacterial functions, it has been considered dispensable for infection. In this issue, Nair et al. revisit this dogma and demonstrate that pAtC58 increases Agrobacterium virulence. Surprisingly, however, this increase was likely due to the enhancement of the vir gene expression rather than to the presence of the attR gene on the pATC58 plasmid.

Although the bacterial genes participating in Agrobacterium attachment to the host plants are relatively well characterized, the involvement of host factors remains largely obscure. Recent progress in this direction stemmed from identification of T-DNA-tagged Arabidopsis mutants defective in their ability to bind Agrobacterium. In this issue, Zhu et al. (2003) report characterization of one such mutant, rat4, which contains a T-DNA insertion in the 3′-untranslated region of cellulose synthase-like gene CSLA9. Their data indicate that CSLA9 is involved in development and growth of lateral roots, determination of sugar composition of plant cell walls, and the ability of the roots to bind Agrobacterium. Furthermore, the CSLA9 promoter exhibited enhanced expression in the root elongation, previously shown to be most susceptible to Agrobacterium-mediated transformation. Similarly, preferential expression in the elongation zones of Arabidopsis roots has been reported for a plant histone gene, H2A-1, which is required for T-DNA integration.

Cell-to-Cell Transport

T strands with their cognate VirD2 are exported into the host cell by a type IV secretion system, which in Agrobacterium is assembled from proteins encoded by the virD4 gene and virB operon, with 11 open reading frames. Interestingly, this system also exports other Agrobacterium proteins, such as VirE3, VirF, and VirE2. Although VirE2 most likely packages the T strand into the T complex, this binding is thought to occur within the cytoplasm of the host cell after independent export of the T strands and VirE2. But what prevents VirE2 from binding to the T strands already within

Agrobacterium? This role has been assigned to the VirE1 chaperone protein that associates with VirE2 and blocks its binding to ssDNA. Furthermore, VirE1 has been suggested also to participate in the VirE2 export. However, convincing evidence to the contrary is presented by Vergunst. Using the Cre Recombinase Reporter Assay for Translocation to study protein export from Agrobacterium into plant and yeast cells, they demonstrate that recognition of VirE2 by the bacterial export machinery and its subsequent translocation into host cells does not depend on the presence of VirE1. Cre Recombinase Reporter Assay for Translocation was then utilized to show that VirE3, another product of the virE locus earlier reported to be exported into the yeast cells, is transferred from Agrobacterium to plants. The function of VirE3 in the plant cell, however, remains completely unknown. Vergunst also localized the Agrobacterium-to-plant export signal to the C-terminal 50 amino acids of VirE2 and VirE3. These and previous studies clearly demonstrate that in addition to DNA, Agrobacterium transfers a variety of its own proteins to the host cell; the challenge to researchers now is to understand how these bacterial proteins participate in the genetic transformation process from within the host cell.

Dna Integration and Expression

T-DNA integration is the culmination point of the entire process of the Agrobacterium-plant cell DNA transfer. But how does a T-DNA molecule insinuate itself into the molecule of the plant genomic DNA? Because the T-DNA does not encode enzymatic activities required for integration, the protein components of the T-complex, i.e. VirD2 and VirE2 and plant DNA ligases, must provide these functions. T-DNA integration was proposed to initiate with ligation of the 5′ end of the T strand to the genomic DNA followed by second strand synthesis by the plant DNA repair machinery. However, another study suggested that the T strand is converted into a double-stranded form before integration.

Obviously, it is the expression of the integrated transgenes that produces tumors in the wild-type infection or desired transgenic phenotypes in genetic engineering experiments. However, some transgenes, although stably integrated in the plant genome, often are not expressed due to their posttranscriptional gene silencing (PTGS), which is characterized by a reduction in transcript levels without affecting the rate of transcription. Furthermore, PTGS of a transgene can also silence its endogenous cellular homologs, resulting in co-suppression. Agrobacterium-mediated gene delivery has served as a valuable experimental tool to study the mechanisms of induction and cell-to-cell spread of PTGS. Continuing this trend, Lee utilized silencing of Agrobacterium oncogenes contained within the T-DNA to demonstrate that transgene sequences influence the effectiveness of PTGS and that sequences required for oncogene silencing must include a translation start site. Unexpectedly, unlike several other cases of PTGS, oncogene silencing was not graft transmissible. In addition to helping us better understand the mechanisms of PTGS, silencing of Agrobacterium oncogenes represents a novel approach to control the crown gall disease, which affects such agronomically important plants as grape, rose, apple, cherry, and others.

Finally, Agrobacterium infection was also used as a model system to study cellular processes required for establishment of plant tumors. Wächter showed that vascular differentiation and disruption of epidermal cell layers play key roles in tumor formation, allowing delivery of water and Suc to the proliferating tumor cells. This work suggested that Agrobacterium-induced tumors represent nutritional sinks to which a high-volume flow of solutes with essential inorganic nutrients is directed in an ethylene-dependent fashion, followed by a Suc-to-hexose shift in sugar balance caused by vacuolar invertases within the growing tumors.

Examples of Genetic Modification

Bananas

In many countries around the world bananas are the main source of calories. According to reports from Uganda, their production is compromised by the emergence of new diseases. Ugandan scientists have successfully used a genetic modification, inserting a pepper gene into bananas, which prevents the fruit from getting the disease.

Onions that do not Make you Cry

In 2008, a New Zealand research team lead by Colin Eady produced an onion that does not make you cry while cutting it. Interestingly, the insertion of a single gene which down regulates the activity of the onion enzyme that makes your eyes water has managed to achieve two things: firstly, onions no longer make your eyes water, and secondly, they now have even more health beneficial sulphur-containing substances than regular onions.

Golden Rice

On 31st July 2000, Ingo Potrykus appeared on the cover of Time magazine. The Swiss scientist and his German colleague Peter Beyerhad had produced a breed of rice which, unlike any other, also contains provitamin A. The lack of this vitamin is especially harmful to the poorest and is estimated to cause blindness among 250,000-500,000 children every year. Another two million people a year die from other deficiency-related causes. So far, the measures taken to introduce vitamin supplements have not yet reached those poor countries. Because of its colour, the product was first given the name golden rice, which remains the same to this day. After concluding numerous tests, researchers from the International Rice Research Institute in the Philippines have proven that even small amounts of the rice are sufficient, and that it is absolutely safe. Unfortunately, partly due to the vandalism of "green" activists, the rice has yet to reach its target group. We should also mention the existence of genetically modified rice, produced in 2011, which contains four times as much iron as the regular one, and could therefore save even more lives.

Purple Tomatoes

In 2008, a small English research group published a study describing how they had transferred a gene from a decorative plant into a tomato, which enabled the production of anthocyanin, making the tomato dark blue. They later tested it on mice, and discovered that it prevents them from getting cancer. These tomatoes cannot be bought due to their GMO status but you can buy similar dark blue ones, which are the result of a complex interspecific hybridisation performed by Italian scientists. Producing such tomatoes seems to have sparked competition between cultivators. In 2012 for instance, the Israelis announced that they have beaten the Italians by introducing a new species to the market called Black Galaxy.

Carrots that Help Prevent Osteoporosis

In 2004, an American research team transferred a CAX1 gene from the mouseear cress into carrots so that they contained larger amounts of organically bound calcium. In 2008, they performed a study where such carrots were tested on mice and 30 volunteers, and the results showed that humans absorbed 42% more calcium from the modified carrots than from regular ones. The aim of this test was to help prevent osteoporosis, while the emphasis was on its bioavailability in target tissues.

Soybean Oil for Frying

Two American companies have significantly improved soybean edible oil through genetic modification. The Plenish oil marketed by DuPont used gene silencing to produce oil that contains low levels of polyunsaturated fats and high levels of monounsaturated fats, while the saturated fatty acids have decreased by 20%. This kind of oil has high stability during baking, meaning it does not need to go through the chemical process of hydrogenation, which produces unwanted trans fats. There is a similar vascular-system-friendly oil marketed by Monsanto with two inserted genes enabling the production of omega-3 fatty acids. Rapeseed is also close to being released into production by the company BASF, containing five genes isolated from seaweed. These are the first products since the Flavr Savr tomato in 1994 to be made directly available to the consumer.

Arctic Apple

An apple turns brown if it is cut in half, which is why its slices are often soaked in antioxidants to prevent this. In 2012, Canadian scientists started releasing two popular apple cultivars, Golden Delicious and Granny Smith, and are planning to introduce Gala and Fuji cultivars, all under the brand name Arctic Apple. The non-browning characteristic could already be found in some apple cultivars, like the newest Slovene cultivar called Majda, but none of them has gained strong commercial recognition. The aim of this research is to prevent the most popular cultivars from turning brown. The same principle applies to these apples as to the previously mentioned onions: it is not about adding genes but about downregulating the activity of the existing ones.

References

- G., Halford, Nigel (2012). Genetically modified crops. World Scientific (Firm) (2nd ed.). London: Imperial College Press. ISBN 978-1848168381. OCLC 785724094
- Transgenic-plants-meaning-reasons-and-fundamentals-588: biotechnologynotes.com, Retrieved 12, March 2020
- Bruening, G.; Lyons, J. M. (2000). "The case of the FLAVR SAVR tomato". California Agriculture. 54 (4): 6–7. doi:10.3733/ca.v054n04p6
- Microinjection-into-plant-cells-of-etiolated-seedlings-981244058: labonline.com.au, Retrieved 30, May 2020
- Andrew Pollack (19 June 2013). "Executive at Monsanto wins global food honor". The New York Times. Retrieved 20, June 2020
- Gene-gun-method-3145: maximumyield.com, Retrieved 11, January 2020
- Koornneef, M.; Meinke, D. (2010). "The development of Arabidopsis as a model plant". The Plant Journal. 61 (6): 909–21. doi:10.1111/j.1365-313X.2009.04086.x. PMID 20409266
- Cisgenics-and-its-Application-in-Crop-Improvement-3861: biotecharticles.com, Retrieved 20, April 2020
- Predieri, S. (2001). "Mutation induction and tissue culture in improving fruits". Plant Cell, Tissue and Organ Culture. 64 (2/3): 185–210. doi:10.1023/A:1010623203554
- 10-successful-examples-of-genetic-modification: metinalista.si, Retrieved 28, May 2020

Chapter 4

Water Management

Agricultural water is the water used for the production of food and fibre. It also includes water used in the watering of livestock and irrigation of crops. A few of the topics studied under agricultural water management are the role of irrigation in food production, irrigation scheduling, ditch irrigation, overhead irrigation, etc. This chapter closely examines these key concepts of agricultural water management to provide an extensive understanding of the subject.

Agricultural Water

Agricultural water is water that is used to grow fresh produce and sustain livestock. The use of agricultural water makes it possible to grow fruits and vegetables and raise livestock, which is a main part of our diet. Agricultural water is used for irrigation, pesticide and fertilizer applications, crop cooling (for example, light irrigation), and frost control. According to the United States Geological Survey (USGS), water used for irrigation accounts for nearly 65 percent of the world's freshwater withdrawals excluding thermoelectric power. There are 330 million acres of land used for agricultural purposes in the United States that produce an abundance of food and other products.

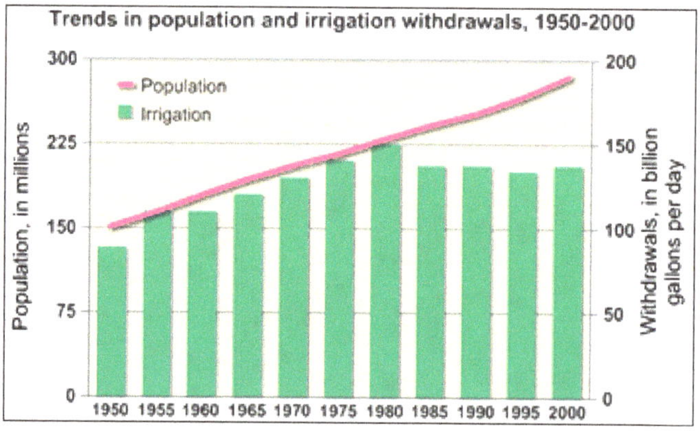

When agricultural water is used effectively and safely, production and crop yield are positively affected. A decrease in applied water can cause production and yield to decrease. Management strategies are the most important way to improve agricultural water use and maintain optimal production and yield. The key is to implement management strategies that improve water use efficiency without decreasing yield. Some examples include improved irrigation scheduling and crop specific irrigation management. These strategies allow for the conservation of water and energy, and decrease grower's costs.

Sources of water used on the farm can be grouped into three types based on the likelihood that they can become contaminated:

(1) Surface water,

(2) Well water,

(3) Municipal water.

Surface water includes ponds, open springs, lakes, rivers, and streams. It has the highest risk for contamination because we often do not have control over what might be entering the water source upstream at any given time. Access of wild and domestic animals, drainage from upstream cattle operations, runoff from manure piles, and sewage discharges are all possible causes for sudden and unexpected surface water contamination.

Water obtained from the wells on your farm generally has an intermediate risk. The potential for well water to become contaminated with harmful microorganisms is greatest when they are located too close to flood zones, septic tanks, cesspools, animal agricultural sites, manure storage areas, or drainage fields. Risks are greatly increased if the wells have not been constructed properly, or if the well casing has become cracked over time. However, if wells are properly sited, constructed, and maintained, they can be a reliable source of contaminant-free water.

Municipal water obtained from your local water authority has the lowest level of food safety risk. We expect this to be the safest type of water because it is required by law to meet the highest chemical and microbiological drinking water standards, and it is tested regularly to ensure that it is consistently safe to drink.

Preventing Agricultural Water from Becoming a Source of Contamination

Postharvest Water: Produce Washing, Handwashing, Cooling, Drinking

- Conduct a potable water test. It is critical to use pathogen-free water for all postharvest water used for washing, flumes and tanks, handwashing, and drinking.
- If postharvest water does not meet the drinking water standard, it may be possible to treat the source with one-time shock chlorination.
- If postharvest water does not meet the drinking water standard, it may be necessary to install a continuous sanitation system using chlorination or ultraviolet (UV) light.

Surface Water used for Irrigation

- Regularly monitor the microbial content of your surface water. Consider testing three times each season:
    - At planting.
    - At peak use.
    - At or near harvest.
- Look for evidence of entry points for animals or areas where runoff can occur. Consider installing fences, vegetative buffer plantings, diversion berms, or other physical structures to protect the water from animal intrusion or drainage from contamination sources.
- When possible, use indirect irrigation methods, such as drip irrigation, that minimize water contact with fruits, tomatoes, peppers, and cole crops.
- Plasticulture methods that cover drip lines provide further protection for lower-growing crops such as leafy greens, cantaloupes, and summer squash.
- Use overhead irrigation in the morning to allow adequate drying of the crop surface before harvest. This speeds the destruction of both human and plant pathogens, and saves water.
- Maximize the time between overhead irrigation and harvest.
- Consider switching to well or municipal water for overhead irrigation and crop spraying methods.

Well Water

- Monitor your well water quality at least twice during the growing season.

- Check that your well is installed correctly. There should be at least 2 inches of grout maintained between the well casing and the surrounding soil to prevent infiltration of surface water. Because well drilling is not regulated in Pennsylvania, your well may not have thc proper casing and grout to exclude surface water contaminants.
- Maintain a 100-foot radius around the well that is kept free from animal intrusion, manure piles, or other contamination sources.
- Install a sanitary well cap to prevent insects or small mammals from entering the well.
- Inspect your wells at least once each year. Check that the well cap and casing seal are in good condition.

Use of Water in Agriculture

Food and agriculture are the largest consumers of water, requiring one hundred times more than we use for personal needs. Up to 70 % of the water we take from rivers and groundwater goes into irrigation, about 10% is used in domestic applications and 20% in industry. Currently, about 3600 km3 of freshwater are withdrawn for human use. Of these, roughly half is really consumed as a result of evaporation, incorporation into crops and transpiration from crops. The other half recharges groundwater or surface flows or is lost in unproductive evaporation. Up to 90% of the water withdrawn for domestic use is returned to rivers and aquifers as wastewater and industries typically consume only about 5% of the water they withdraw. This wastewater from domestic sewage systems and industries should be treated before being dismissed.

Since the 1960s the global nutrition has considerably improved, providing more food per capita at progressively lower prices. This performance was possible through high-yielding seeds, irrigation and plant nutrition. As population keeps increasing more food and livestock feed need to be produced in the future and more water applied to this purpose. Irrigate agriculture will have to claim large quantities of water to produce the food required to feed the world. The main source of food for the population of the world is agriculture: this term also includes livestock husbandry, manages fisheries and forestry.

The composition of meals changes gradually as lifestyles change. What agriculture produces is driven by consumer demand, and changes in consumer preferences have an influence on the water needed for food production.

Cereals are by far the most important source of total food consumption: in developing countries the consumption of cereals 30 years ago represented 61% of total calories. It decreased to 56% nowadays and this reflects diet diversification, proving that more countries achieve higher levels of nutrition. It is expected that cereals will continue to supply more than 50% of the food consumed in the foreseeable future. A large proportion of cereals is produced for animal feed.

Food production from the livestock sector includes meat (beef, pork, poultry, etc.), dairy production and eggs.

For vegetative growth and development plants require water in adequate quantity and at the right time. Crops have very specific water requirements, and these vary depending on local climate conditions. The production of meat requires between six and twenty time more water than for cereals.

The following tables give an overview of the water consumption in food and agriculture.

Specific values for the water equivalent of a selection of food products are given in the first table. The second table shows the amount of water needed necessary for a few products per unit of consumption. A glass of wine acquires for example 120 liters of water, a hamburger 2.400 liters, a cotton shirt 4.000 liters and a couple of shoes made out of cows leather 8.000 liters.

The third table gives an overview of the amount of virtual water used in the different kind of agricultural products. The big difference between the countries is due to the climate, applied technology and the amount of production related to it.

Product	Unit	Equivalent water in m^3 per unit
Cattle	Head	4000
Sheep	Head	500
Fresh beef	Kg	15
Fresh lamb	Kg	10
Fresh poultry	Kg	6
Cereals	Kg	1.5

Citrus fruits	Kg	1
Palm oil	Kg	2
Pulses, rood and tubers	Kg	1

Product	Quantity	Equivalent water in liters
Glass of beer	250 ml	75
Glass of wine	125 ml	120
Glass of milk	200 ml	200
Glass of apple juice	200 ml	190
Cup of coffee	125 ml	140
Glass of orange juice	200 ml	170
Cup of tea	250 ml	35
Chips bag	200 g	185
Slice of bread	30 g	40
Egg	40 g	135
Slice of bread with cheese	30 g+ 10 g	90
Hamburger	150 g	2400
Potato	100 g	25
Tomato	70 g	13
Apple	100 g	70
Orange	100 g	50
Cotton T- shirt	500 g	4100
Pair of shoes	1	8000
Sheet A4 paper	80 g/m²	10
Microchip	2g	32

Amount of virtual water per food per country in m³/ton												
	U.S.	China	India	Rus-sia	Indone-sia	Australia	Brazil	Japan	Mexico	Italy	Nether lands	World average
Rice	1903	1972	4254	3584	3209	1525	4600	1822	3257	2506		3419
Wheat	849	690	1654	2375		1588	1616	734	1066	2421	619	1334
Corn	489	801	1937	1397	1285	744	1180	1493	1744	530	408	909
Soya beans	1869	2617	4124	3933	2030	2106	1076	2326	3177	1506		1789
Sugarcane	103	117	159		164	141	155	120	171			175
Cottonseed	2535	1419	8264		4453	1887	2777		2127			3644
Carton plaxel	5733	3210	18694		10072	4268	6281		4812			8242
Coconut		749	2255		2071		1590		1954			2545
Roast coffee	5790	7488	14500		21030		16633		33475			20682
Tea leaves		11110	7002	3002	9474		6592	4940				9205
Beef	13193	12560	16482	21028	14818	17112	16961	11019	37762	21167	11681	15497
Pork	3946	2211	4397	6947	3938	5909	4818	4962	6559	6377	3790	4856
Goat's meet	3082	3994	5i87	5290	4543	3839	4175	2560	10252	4180	2791	4043

Mutton	5977	5202	6692	7621	5956	6947	6267	3571	16878	7572	5298	6143
chicken	2389	3652	7736	5763	5549	2914	3913	2977	5013	2198	2222	3918
Eggs	1510	3550	7531	4919	5400	1844	3337	1884	4277	1389	1404	3340
Milk	695	1000	1369	1345	1143	915	1001	812	2382	861	641	990
Milk powder	3234	4648	6368	6253	5317	4255	4654	3774	11077	4005	2982	4602
Cheese	3457	4963	6793	6671	5675	4544	4969	4032	11805	4278	3190	4914
Cow leather	14190	13513	17710	22575	15929	18384	18222	11864	40482	22724	12572	16656

The amount of water involved in agriculture is significant and most of it is provided directly by rainfall. A rough calculation of global water needs for food production can be based on the specific water requirements to produce food for one person. The present average food ingest 2800 kcal/person/day may require 1000 m^3 per year to be produced. The world population is about 6 billion, so water needed to produce the necessary food, excluding water losses due to the irrigation system, is 6000 km^3. Most of it is provided by rainfall stored in the soil profile and only 15% is provided through irrigation. Irrigation therefore needs 900 km^3 of water per year for food crops. On average just about 40% of water withdrawn from rivers, lakes and aquifers for agriculture effectively contribute to crop production (the rest is lost through evaporation and deep infiltration). Consequently the current global water withdrawals for irrigation are estimated to be about 2000 to 2500 km^3 per year.

The irrigation level varies from area to area, mostly depending on climate conditions and on the development of irrigation infrastrure. The following figure shows the area equipped for irrigation as percentage of cultivated land by country.

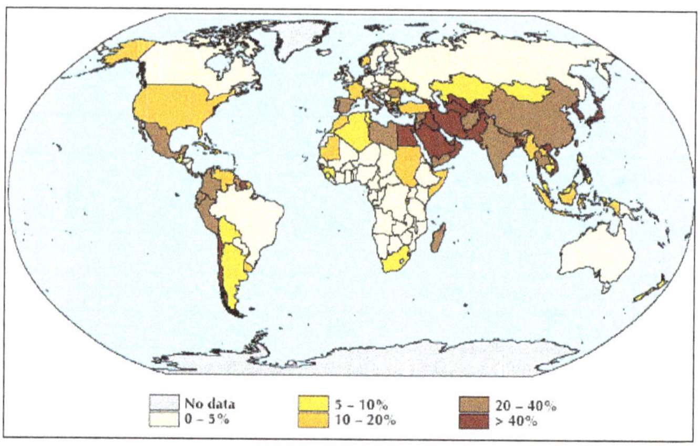

Irrigation-water management has a log way to adapt to the increasing production requirements, however water-saving technologies are already available and can significantly reduce the waste of water. If incentives are in place, as increasing the price of irrigation water, farmers will adopt water-saving irrigation technologies. The main technologies likely to be used in developing countries, where labor is normally abundant but capital scarce, are underground and drip irrigation. Both technologies depend on the frequent application of small amounts of water as directly as possible to the roots of crops. Reducing the pollution loads of water used by farms, industries and urban areas would enable much more of it to be re-used in irrigation. There are enormous potential benefits to be had from the use of wastewater for irrigation.

Agriculture will remain the dominant user of water at the global level. In many countries, in particular those situated in the arid and semi-arid regions of the world, this dependency can be expected to intensify. The contribution of irrigated agriculture to food production is substantial but in future the rate of growth will be lower than in the past. Both irrigated and non-irrigated agriculture still have scope for increasing productivity, including water productivity. Arguably, the expansion of irrigated agriculture protected people on the nutritional fringe from premature death, and preserved tracts of land under forest and wetlands from encroachment by hard-pressed farmers.

Role of Irrigation in Food Production

For vegetative growth and development, plants require, within reach of their roots, water of adequate quality, in appropriate quantity and at the right time. Most of the water a plant absorbs performs the function of raising dissolved nutrients from the soil to the aerial organs, from where it is released to the atmosphere by transpiration: agricultural water use is intrinsically consumptive. Crops have specific water requirements, and these vary depending on local climatic conditions. Whereas an indicative figure for producing one kilogram of wheat is about 1000 litres of water that is returned to the atmosphere, paddy rice may require twice this amount. The production of meat requires between six and twenty times more water than for cereals, depending on the feed/meat conversion factor. Specific values for the water equivalent of a selection of food products are given in table below. Water required for human food intake can be derived from these specific values in a grossly approximate way, depending on the size and composition of the meals.

Water requirement equivalent of main food products.

Product	Unit	Equivalent water in m^3 per unit	
Cattle	head	4000	
Sheep and goats	Head	500	This table gives examples of water required per unit of major food products.
Fresh beef	Kg	10	Including livestock, which consume the most water per unit.
Fresh lamb	Kg	10	Cereals, oil crops, and pulses, roots and tubers consume far less water.
Fresh poultry	Kg	6	
Cereals	Kg	1.5	
Citrus fruits	Kg	1	
Palm oil	Kg	2	
Pulses, root and tubers	Kg	1	

This table gives examples of water required per unit of major food products, including livestock, which consume the most water per unit. Cereals, oil crops, and pulses, roots and tubers consume far less water.

Food Production

Dominant Role of Rainfed Agriculture

Non-irrigated (rainfed) agriculture depends entirely on rainfall stored in the soil profile. This form of agriculture is possible only in regions where rainfall distribution ensures continuing availability of soil moisture during the critical growing periods for the crops. Non-irrigated agriculture accounts for some 60 percent of production in the developing countries. In rainfed agriculture, land management can have a significant impact on crop yields: proper land preparation leading surface runoff to infiltrate close to the roots improves the conservation of moisture in the soil. Various forms of rainwater harvesting can help to retain water in situ. Rainwater harvesting not only provides more water for the crop but can also add to groundwater recharge and help to reduce soil erosion. Other methods are based on collecting water from the local catchment and either relying on storage within the soil profile or else local storage behind bunds or ponds and other structures for use during dry periods. Recently, conservation agriculture practices such as conservation tillage have proven to be effective in improving soil moisture conservation.

The potential to improve non-irrigated yields is restricted where rainfall is subject to large seasonal and interannual variations. With a high risk of yield reductions or complete loss of crop from dry spells and droughts, farmers are reluctant to invest in inputs such as plant nutrients, high-yielding seeds and pest management. For resource-poor farmers in semi-arid regions, the overriding requirement is to harvest sufficient food stuff to ensure nutrition of the household through to the next harvest. This objective may be reached with robust, drought-resistant varieties associated with low yields. Genetic engineering has not yet delivered high-yield drought-resistant varieties, a difficult task to achieve because, for most crop plants, drought resistance is associated with low yields.

Role of Irrigation in Food Production

In irrigated agriculture, water taken up by crops is partly or totally provided through human intervention. Irrigation water is withdrawn from a water source (river, lake or aquifer) and led to the field through an appropriate conveyance infrastructure. To satisfy their water requirements, irrigated crops benefit from both more or less unreliable natural rainfall, and from irrigation water. Irrigation provides a powerful management tool against the vagaries of rainfall and makes it economically attractive to grow high-yield seed varieties and to apply adequate plant nutrition as well as pest control and other inputs, thus giving room for a boost in yields. Figure below illustrates the typical yield response of a cereal crop to water availability and the synergy between irrigation, crop variety and inputs. Irrigation is crucial to the world's food supplies. In 1998, irrigated land made up about one-fifth of the total arable area in developing countries but produced two-fifths of all crops and close to three-fifths of cereal production.

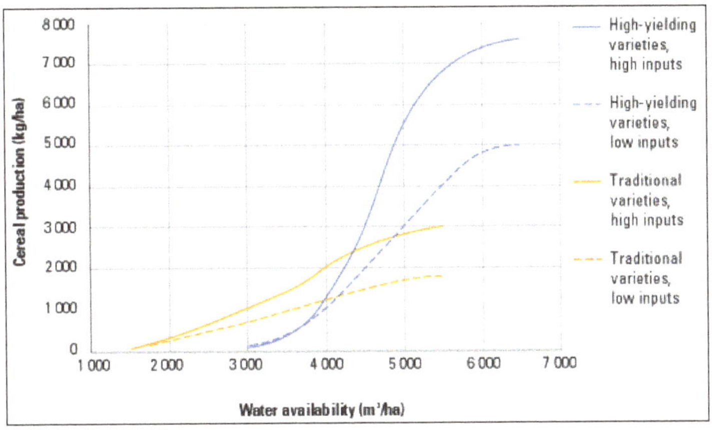

The developed countries account for a quarter of the world's irrigated area (67 million ha). Their annual growth of irrigated area reached a peak of 3 percent in the 1970s and dropped to only 0.2 percent in the 1990s. The population of this group of countries is growing only slowly and therefore a very slow growth in their demand and production of agricultural commodities is foreseen. The focus of irrigation development is consequently expected to be concentrated on the group of developing countries where demographic growth is strong. Increasing competition from the higher valued industrial and domestic sector results in a decrease in the amount of overall water allocated to irrigation. Figure below illustrates the case for the Zhanghe irrigation system in China.

Map below shows irrigated land as percentage of arable land in developing countries. A high proportion of irrigated land is usually found in countries and regions with an arid or semi-arid climate. However, low proportions of irrigated land in sub-Saharan Africa point also to underdeveloped irrigation infrastructure. Data and projections of irrigated land compared to irrigation potential in developing countries are shown in figure below. The irrigation potential figure already takes into account the availability of water. The graph shows that a sizeable part of irrigation potential is already used in the Near East/North Africa region (where water is the limiting factor) and in Asia (where land is often the limiting factor), whereas a large potential is still unused in sub-Saharan Africa and in Latin America.

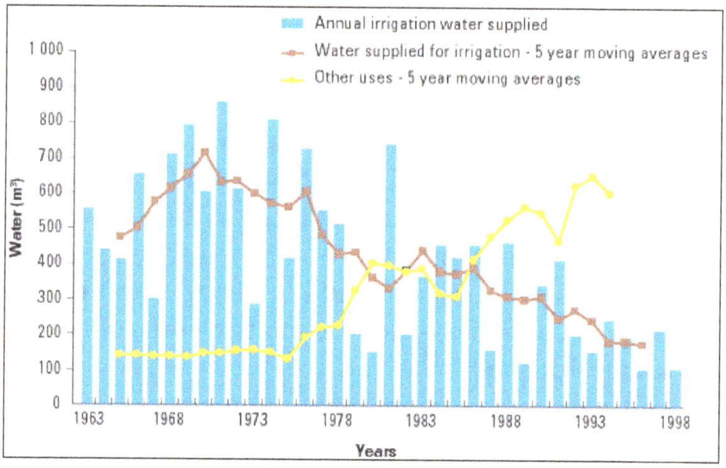

Competing uses of water.

According to FAO forecasts, the share of irrigation in world crop production is expected to increase in the next decades. In particular in developing countries, the area equipped for irrigation is expected to have expanded by 20 percent (40 million ha) by 2030. This means that 20 percent of total land with irrigation potential but not yet equipped will be brought under irrigation, and that 60 percent of all land with irrigation potential (402 million ha) will be in use by 2030. The net increase in irrigated land (40 million ha, 0.6 percent per year) projected to 2030 is less than half the increase over the preceding 36 years (99 million ha, 1.9 percent per year). The projected slowdown in irrigation development reflects the projected lower growth rate of food demand, combined with the increasing scarcity of suitable areas for irrigation and of water resources in some countries, as well as the rising cost of irrigation investment. The first selection of economically attractive irrigation projects has already been implemented, and prices for agricultural commodities have not risen to encourage investment in a second selection of more expensive irrigation projects.

Most of the expansion in irrigated land is achieved by converting land in use in rainfed agriculture or land with rainfed production potential but not yet in use into irrigated land. The expansion of irrigation is projected to be strongest in South Asia, East Asia and Near East/North Africa. These regions have limited or no potential for expansion of non-irrigated agriculture. Arable land expansion will nevertheless remain an important factor in crop production growth in many countries in sub-Saharan Africa, Latin America and some countries in East Asia, although to a much smaller extent than in the past. The growth in wheat and rice production in the developing countries will increasingly come from gains in yield, while expansion of harvested land will continue to be a major contributor to the growth in production of maize.

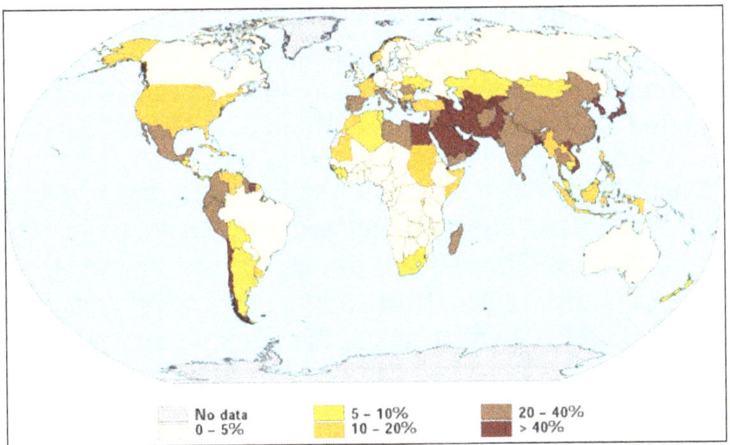

Area equipped for irrigation as percentage of cultivated land by country.

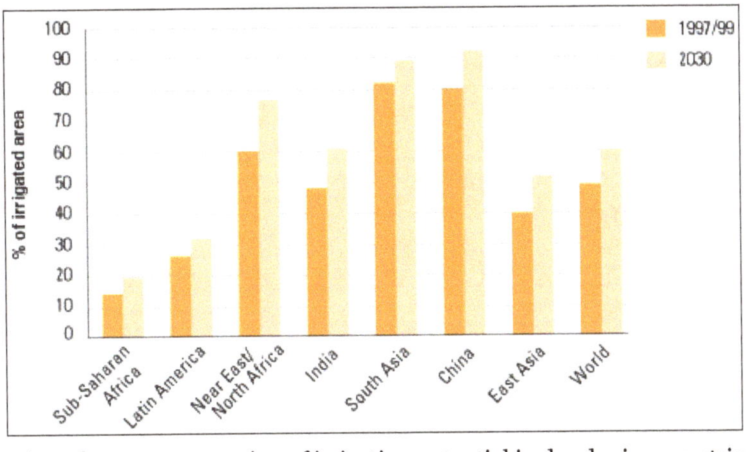

Irrigated area as proportion of irrigation potential in developing countries.

Future Investments in Irrigation

In many developing countries, investments in irrigated infrastructures have represented a significant share of the overall agricultural budget during the second half of the twentieth century. The unit cost of irrigation development varies with countries and types of irrigated infrastructures, ranging typically from US$1 000 to US$10 000 per hectare, with extreme cases reaching US$25 000 per hectare (these costs do not include the cost of water storage as the cost of dam construction varies on a case-by-case basis). The lowest investment costs in irrigation are in Asia, which has the bulk of irrigation and where scale economies are possible. The most expensive irrigation schemes are found in sub-Saharan Africa, where irrigation systems are usually smaller and developing land and water resources is costly.

In the future, the estimates of expansion in land under irrigation will represent an annual investment of about US$5 billion, but most investment in irrigation, between US$10 and 12 billion per year, will certainly come from the needed rehabilitation and modernization of aging irrigated schemes built during the years 1960-1980. In the 1990s, annual investment in storage for irrigation was estimated at about US$12 billion (WCD, 2000). In the future, the contrasting effects of reduced demand for irrigation expansion and increased unit cost of water storage will result in an annual investment estimated between US$4 and 7 billion in the next thirty years.

Typically, investment figures in irrigation do not include that part of the investment provided by the farmer in land improvement and on-farm irrigation that can represent up to 50 percent of the overall investment. In total, it is estimated that annual investment in irrigated agriculture will therefore range between US$25 and 30 billion, about 15 percent of annual expected investments in the water sector.

Water use Efficiency

Assessing the impact of irrigation on available water resources requires an estimate of total abstraction for the purpose of irrigation from rivers, lakes and aquifers. The volume extracted is considerably greater than the consumptive use for irrigation because of conveyance losses from the withdrawal site to the plant root zone. Water use efficiency is an indicator often used to express the level of performance of irrigation systems from the source to the crop: it is the ratio between estimated plant requirements and the actual water withdrawal.

On average, it is estimated that overall water use efficiency of irrigation in developing countries is about 38 percent. Map below shows the importance of agriculture in the countries' water balance, and figure above shows the expected growth in water abstraction for irrigation from 1998 to 2030. The predictions are based on assumptions about possible improvements in irrigation efficiency in each region. These assumptions take into account that, from the farmer's perspective, wherever water is abundant and its cost low, the incentives to save water are limited. Conversely, if farmers can profitably irrigate more land using their allocation in an optimum way, irrigation efficiency may reach higher levels.

Improving irrigation efficiency is a slow and difficult process that depends in large part on the local water scarcity situation. It may be expensive and requires willingness, know-how and action at various levels. Table below shows current and expected water use efficiency for developing countries in 1998 and 2030, as estimated by FAO. The investment and management decisions leading to higher irrigation efficiency are taken and involve irrigation system management and the system-dependent farmers. National water policy may encourage water savings in water-scarce areas by providing incentives and effectively enforcing penalties. When upstream managers cannot ensure conveyance efficiency, there may be no incentives for water users to make efficiency gains. With groundwater, this caveat may not apply since the incentive is generally internalized by the users, and in many cases groundwater users show much greater efficiency than those depending on surface resources. Provides an overview of different aspects of potential improvements in agricultural water use efficiency.

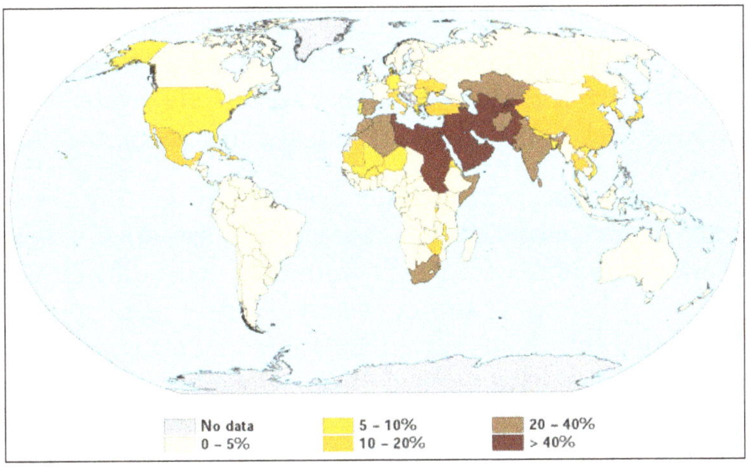

Agricultural water withdrawals as percentage of renewable water resources.

Future Water Withdrawals for Irrigation

Irrigation water withdrawal in developing countries is expected to grow by about 14 percent from the current 2 130 km3 per year to 2 420 km3 in 2030. This finding is based specifically on individual assessments for each developing country. Harvested irrigated area (the cumulated area of all crops during a year) is expected to increase by 33 percent from 257 million ha in 1998 to 341 million ha in 2030. The disproportionate increase in harvested area is explained by expected improvements in irrigation efficiency, which will result in a reduction in gross irrigation water abstraction per ha of crop. A small part of the reduction is due to changes in cropping patterns in China, where consumer preference is causing a shift from rice to wheat production.

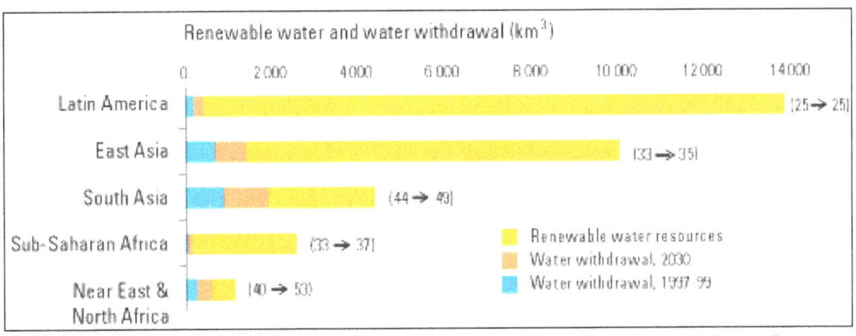

Irrigation and water resources: current and predicted withdrawals.

	Sub-Saharan Africa	Latin America	Near East & North Africa	South Asia	East Asia	All countries
Water use efficiency in irrigation (%)						
1998	33	25	40	44	33	38
2030	37	25	53	49	35	42
Irrigation water withdrawal as a percentage of renewable water resources (%)						
1998	2	1	53	36	8	8
2030	3	2	58	41	8	9

Water use efficiency in 1998 and 2030 (predicted) in 93 developing countries.

While some countries have reached extreme levels of water use for agriculture, irrigation still represents a relatively small part of total water resources of the developing countries. The projected increase in water withdrawal will not significantly alter the overall picture. At the local level, however, there are already severe water shortages, in particular in the Near East / North Africa region and in large parts of Asia.

Of the ninety-three developing countries surveyed by FAO, ten are already using more than 40 percent of their renewable water resources for irrigation, a threshold used to flag the level at which countries are usually forced to make difficult choices between their agricultural and urban water supply sectors. Another eight countries were using more than 20 percent, a threshold that can be used to indicate impending water scarcity. By 2030 South Asia will have reached the 40 percent level, and the Near East and North Africa not less than 58 percent. However, the proportion of renewable water resources allocated to irrigation in sub-Saharan Africa, Latin America and East Asia in 2030 is likely to remain far below the critical threshold.

Special Role of Groundwater

Water contained in shallow underground aquifers has played a significant role in developing and diversifying agricultural production. This is understandable from a resource management perspective: when groundwater is accessible it offers a primary buffer against the vagaries of climate and surface water delivery. But its advantages are also quite subtle. Access to groundwater can occasion a large degree of distributive equity, and for many farmers, groundwater has proved to be a perfect delivery system. Because groundwater is on demand and just-in-time, farmers have sometimes made private investments in groundwater technology as a substitute for unreliable or inequitable surface irrigation services. In many senses, groundwater has been used by farmers to break out of conventional command and control irrigation administration. Some of the management challenges posed by large surface irrigation schemes are avoided, but the aggregate impact of a large number of individual users can be damaging, and moderating the 'race to the pump-house' has proved difficult. However, as groundwater pumping involves a direct cost to the farmer, the incentives to use groundwater efficiently are high. These incentives do not apply so effectively where energy costs are subsidized; such distortion has arguably accelerated groundwater depletion in parts of India and Pakistan.

The technical principles involved in sustainable groundwater and aquifer management are well known but practical implementation of groundwater management has encountered serious difficulties. This is largely due to groundwater's traditional legal status as part of land property and the competing interests of farmers withdrawing water from common-property aquifers. Abstraction can result in water levels declining beyond the economic reach of pumping technology; this may penalize poorer farmers and result in areas being taken out of agricultural production. When near the sea, or in proximity to saline groundwater, over-pumped aquifers are prone to saline intrusion. Groundwater quality is also threatened by the application of fertilizers, herbicides and pesticides that percolate into aquifers. These 'non-point' sources of pollution from agricultural activity often take time to become apparent, but their effects can be long-lasting, particularly in the case of persistent organic pollutants.

Fossil groundwater, that is, groundwater contained in aquifers that are not actively recharged, represent a valuable but exhaustible resource. Thus, for example, the large sedimentary aquifers of North Africa and the Middle East, decoupled from contemporary recharge, have already been exploited for large-scale agricultural development in a process of planned depletion. The degree to which further abstractions occur will be limited in some cases by the economic limits to pumping, and promoted where strong economic demand from agriculture or urban water supply becomes effective. Two countries, Libyan Arab Jamahiriya and Saudi Arabia, are already using considerably more water for irrigation than their annual renewable resources, by drawing on fossil groundwater reserves. Several other countries rely to a limited extent on fossil groundwater for irrigation. Where such groundwater reserves have a high strategic value in terms of water security, the depletion of such reserves to irrigate is questionable.

Role of Irrigation in Alleviating Poverty and Improving Food Security

There is a positive, albeit complex, link between water services for irrigation and other farm use, poverty alleviation and food security. Many of the rural poor work directly in agriculture, as smallholders, farm laborers or herders. The overall impact can be remarkable: in India, for example, in unirrigated districts 69 percent of people are poor, while in irrigated districts, only 26 percent are poor. Their income can be boosted by pro-poor measures, such as ensuring fair access to land, water and other assets and inputs, and to services, including education and health. Relevant reforms of agricultural policy and practices can strengthen these measures.

The availability of water confers opportunities to individuals and communities to boost food production, both in quantity and diversity, to satisfy their own needs and also to generate income from surpluses. Irrigation has a land-augmenting effect and can therefore mean the difference between extreme poverty and the satisfaction of the household's basic needs. It is generally recognized that in order to have an impact on food security, irrigation projects need to be integrated with an entire range of complementary measures, ranging from credit, marketing and agricultural extension advice to improvement of communications, health and education infrastructure. Land tenure may also represent a major constraint: irrigation schemes controlled by absentee landowners and serving distant markets, even when highly efficient, may fail to improve local food security when both commodities and benefits are exported.

Irrigation projects are as diverse as the local situations in which they are implemented. Generally, small-scale irrigation projects including projects based on shallow groundwater pumping provide a manageable framework that can give control to the local poor and avoid leaking resources to the non-poor. Large-scale irrigation, as may be determined by the need to carry out large-scale engineering works to harness water and convey it to the fields, can also be made to work for the poor provided that the benefits can be shared equitably, and investment, operation and maintenance costs are efficiently covered.

Managing Agricultural Risk for Sustainable Livelihoods

Community-managed small-scale irrigation systems, by improving yields and cropping intensities, have proved effective in alleviating rural poverty and eradicating food insecurity. Marketing of agricultural produce, both locally and at more distant places when adequate transport and communication infrastructure is available, can make a significant contribution to the income of farmers. Bank deposits and credits, as well as crop insurance, can be used to finance agricultural operations and buffer against climatic risk. However, banking services usually are not accessible to people who have no collateral assets. Many rural credit systems may also not accommodate pay-back over a period of years – the time needed to realize the benefits from investment in irrigation technology. However, non-conventional credit systems based on trust and social solidarity can support poor farmers. Improvement in poor people's food commodity storage facilities reduces post-harvest losses and can save significant amounts of food, thus contributing to food security. Similarly, where technically and financially possible, storing water in surface reservoirs and aquifers is a strategic means for managing agricultural risk. Water in the reservoir or the aquifer is, in a way, equivalent to money in the bank.

Irrigation Contributes to Creating Off-farm Employment

Irrigation, supported through inputs such as high-yielding varieties, nutrients and pest management, together with a more extended agricultural season, higher cropping intensity and a more diverse assortment of crops, can generate rural employment in other non-agricultural services. The productivity boost provided by irrigated agriculture results in increased and sustained rural employment, thereby reducing the hardships experienced by rural populations that might otherwise drift to urban areas under economic pressure. Growth in the incomes of farmers and farm laborers creates increased demand for basic non-farm products and services in rural areas. These goods and services are often difficult to trade over long distances. They tend to be produced and provided locally, usually with labour-intensive methods, and so have great potential to create employment and alleviate poverty. Studies in many countries have shown multipliers ranging from two (in Malaysia, India and the United States) to six.

The Contribution of Fisheries and Aquaculture to Food Security

Fish has a very good nutrient profile and is an excellent source of high-quality animal protein and of highly digestible energy. Refugees and displaced people facing food insecurity may turn to fishing for survival, where this possibility exists. Staple-food fish, often low-valued fish species, is in high demand in most developing countries, because of its affordability.

Inland fish production provides significant contributions to animal protein supplies in many rural areas. In some regions freshwater fish represent an essential, often irreplaceable source of cheap high-quality animal protein, crucial to a well-balanced diet in marginally food-secure communities. Most inland fish produce is consumed locally, marketed domestically, and often contributes to the subsistence and livelihood of poor people. The degree of participation in fishing and fish farming is high in many rural communities. Fish production is often undertaken in addition to agricultural or other activities. Yields from inland capture fisheries, especially subsistence fisheries, can be very significant, even though they are often greatly under-reported. Yields from inland capture fisheries are highest in Asia in terms of total volumes, but are also important in sub-Saharan Africa. Fishery enhancement techniques, especially stocking of natural and artificial water bodies, are making a major contribution to the total catch.

Rural aquaculture contributes to the alleviation of poverty directly through small-scale household farming of aquatic organisms for domestic consumption and income. It also contributes indirectly through employment of the poor as service providers to aquaculture or as workers on aquatic farms. Poor rural and urban consumers can greatly benefit from low-cost fish provided by aquaculture. For aquaculture to be effective in alleviating poverty, it should focus on low-cost products favored by the poor, and emphasis should be placed on aquatic species feeding low in the food chain. The potential exists for aquaculture production for local markets and consumers. Combined rice-fish systems are possible and carry high benefits because they provide cereals and protein at the same time. These systems have also been shown to have beneficial effects on the malaria situation, where mosquito vectors breed in rice fields and where the selected fish species feed on the mosquito larvae.

Forestry and Food Security

A large number of forest products contribute to food security: FAO estimates that about 1.6 billion people in the world rely to a certain extent on forest resources for their livelihoods. For most rural people in the world, firewood provides the fuel for cooking food, and so its availability is an integral part of food security. The bio-energy sector generates employment and income in developing countries.

Most forests and tree plantations subsist on rainwater, or may develop around irrigation schemes. Some tree species can use large amounts of water drawn from water storage in the soil profile and in shallow aquifers. Trees in forests and outside of forests provide significant benefits to poor people and contribute to food security. The benefits from water used by forests are largely seen in terms of wood and non-wood products, as well as in protecting the environment, reducing soil degradation, and preserving biodiversity.

Irrigation Scheduling

Scheduling irrigation is attempting to apply water to potatoes at the appropriate amount for a specific stage in the plants development and growth. The potato plant's use of water is primarily for transpiration and tuber production and, therefore, irrigation is most important from emergence to vine senescence. Transpiration is the movement of water through the vine, from roots to leaves, to compensate for water loss at stomates (pores) that are open to allow gas exchange (oxygen and carbon dioxide) between the plant and the atmosphere thereby supporting plant photosynthesis and respiration.

Principles of Irrigation Scheduling from Soil-water View

1. Soil moisture. Sites for monitoring soil moisture should be chosen to be most representative of the field. The purpose is to limit under-watering of the heavier soils and over-watering lighter soils. For precision irrigation where watering can be controlled in smaller areas within the field, more monitors would be needed and both better and poorer soils would need to be monitored.

2. Root zone depth. Root zone depth is the zone where most of the root structure is found. This varies with different potato varieties but as a general rule, roots develop down to 18 inches below the seed piece.

3. Water holding parameters. Two measurements would be important. The "fill point" is the wettest a soil can be before water drains below the root zone. This

would be near 100% field capacity (FC) or 100% holding capacity of the root zone and depends on soil texture. In general, sandy soils have the lowest FC while silt loams have the highest with clays being intermediate. The "refill point" is the driest a soil can be before daily water use is lowered due to too little water in the root zone. This begins to induce the shutting of stomates resulting in reduction of carbohydrate synthesis (photosynthesis) and respiration (metabolism), and leads to wilting. This has a direct relation to yield. The difference between field capacity and 40% depreciation is the "allowable depletion" (AD) amount of water and, for potato, is 20-25% FC or about half the total available water (about 40% FC). In sandy loam soils, the AD is three-quarters to one-inch water up to a depth of 12 inches or one to one and a half inch for the root zone of a full-grown determinate potato variety. Soils that are compacted or tend to seal will lower water-holding capacity and reduce penetration of water into the soil.

4. Effective irrigation. Effective irrigation is the amount of water that actually gets into the root zone and is available to the plant. Some of the irrigation water (actual irrigation) is lost as run-off, evaporation or deep percolation.

5. Daily water use. Daily water usage by the potato is dependent on the growth stage of the plant and environmental conditions on that day. It is directly related to canopy development, mostly leaves which contain nearly all the stomates. Environmental conditions that affect daily water use are air temperature, relative humidity, wind, and solar radiation. An excellent guide to daily water usage is evapotranspiration (ET) data that is calculated from weather station data. Most, if not all, land-grant universities like University of Nebraska calculate ETs for stations around the state and provide them on the web and for publication in local newspapers. ET is the total daily water use from both transpiration by the plant and evaporation from the soil and can be as high as a third of an inch in a day for potato. ET needs also to be adjusted for canopy size or row closure, canopy width divided by row spacing; 80% row closure is considered full canopy or 100% ET. Therefore, a potato crop that has row-closed on sandy loam at field capacity (three-quarters to one inch AD) can be carried for two and a half to three days under high ET conditions. From this, one can estimate a two to two and a half inches weekly irrigation requirement for potatoes during tuber bulking under high ET conditionshigh temperature, low relative humidity, intense sunlight, high winds, and long days. Seasonal ETs differ for crops due to duration at full canopy and growing season.

Key Factors in Managing Irrigation

- How much water gets into the soil,
- How much water the soil can hold,
- How much water is being used by the plant.

The accurate determination of an irrigation schedule is a time-consuming and complicated process. The introduction of computer programs, however, has made it easier and it is possible to schedule the irrigation water supply exactly according to the water needs of the crops. Ideally, at the beginning of the growing season, the amount of water given per irrigation application, also called the irrigation depth, is small and given frequently. This is due to the low evapotranspiration of the young plants and their shallow root depth. During the mid season, the irrigation depth should be larger and given less frequently due to high evapotranspiration and maximum root depth. Thus, ideally, the irrigation depth and/or the irrigation interval vary with the crop development.

When sprinkler and drip irrigation methods are used, it may be possible and practical to vary both the irrigation depth and interval during the growing season. With these methods it is just a matter of turning on the tap longer/shorter or less/more frequently.

When surface irrigation methods are used, however, it is not very practical to vary the irrigation depth and frequency too much. With, in particular, surface irrigation, variations in irrigation depth are only possible within limits. It is also very confusing for the farmers to change the schedule all the time. Therefore, it is often sufficient to estimate or roughly calculate the irrigation schedule and to fix the most suitable depth and interval; in other words, to keep the irrigation depth and the interval constant over the growing season. Three simple methods to determine the irrigation schedule are briefly described: plant observation method, estimation method and simple calculation method.

The plant observation method is the method which is normally used by farmers in the field to estimate "when" to irrigate. The method is based on observing changes in plant characteristics, such as changes in colour of the plants, curling of the leaves and ultimately plant wilting.

The simple calculation method is based on the estimated depth (in mm) of the irrigation application, and the calculated irrigation water need of the crop during the growing season.

Plant Observation Method

The plant observation method determines "when" the plants have to be irrigated and is based on observing changes in the plant characteristics, such as changes in colour of the plants, curling of the leaves and ultimately plant wilting. The changes can often only be detected by looking at the crop as a whole rather than at the individual plants. When the crop comes under water stress the appearance changes from vigorous growth (many young leaves which are light green) to slow or even no growth (fewer young leaves, darker in colour, and sometimes greyish and dull).

Some crops react to water stress by changing their leaf orientation: with adequate water available, the leaves are perpendicular to the sun. However, when little water is available, the leaves turn away from the sun (thus reducing the transpiration and production).

To use the plant observation method successfully, experience is required as well as a good knowledge of the local circumstances. A farmer will, for example, know where the sandy spots in the field are, which is where the plants will first show stress characteristics: the colour changes and wilting are more pronounced on the sandy spots.

The disadvantage of the plant observation method is that by the time the symptoms are evident, the irrigation water has already been withheld too long for most crops and yield losses are already inevitable. It is important to note that it is not advisable to wait for the symptoms. Especially in the early stages of crop growth, irrigation water has to be applied before the symptoms are evident.

Another indicator of water availability is the leaf temperature. If the leaves are cool during the hot part of the day, the plants do not suffer from water stress. However, if the leaves are warm, irrigation is needed. Special devices have been developed to measure the leaf temperature in relation to the air temperature. However, they must be calibrated for specific conditions before being used to determine the irrigation schedule.

Another method used to determine the irrigation schedule involves soil moisture measurements in the field. When the soil moisture content has dropped to a certain critical level, irrigation water is applied. Instruments to measure the soil moisture include gypsum blocks, tensiometers and neutron probes. Their use, however, is beyond the scope of this manual.

Estimation Method

Estimating the Irrigation Schedule

Here, a table is provided to estimate the irrigation schedule for the major field crops during the period of peak water demand; the schedules are given for three different soil types and three different climates. The table is based on calculated crop water needs and an estimated root depth for each of the crops under consideration. The table assumes that with the irrigation method used the maximum possible net application depth is 70 mm.

With respect to soil types, a distinction has been made between sand, loam, and clay, which have, respectively, a low, a medium and high available water content. With respect to climate, a distinction is made between three different climates.

Shallow and/ or sandy soil	In a sandy soil or a shallow soil (with a hard pan or impermeable layer close to the soil surface), little water can be stored; irrigation will thus have to take place frequently but little water is given per application
Loamy soil	In a loamy soil more water can be stored than in a sandy or shallow soil. Irrigation water is applied less frequently and more water is given per application.
Clayey soil	In a clayey soil even more water can be stored than in a medium soil. Irrigation water is applied even less frequently and again more water is given per application.
Climate 1	Represents a situation where the reference crop evapotranspiration ETo = 4 - 5 mm/day.
Climate 2	Represents an ETo = 6 - 7 mm/day.
Climate 3	Represents an ETo = 8 - 9 mm/day.

An overview indicating in which climatic zones these ETo values can be found is given below:

Climatic zone	Mean daily temperature		
	low (less than 15°C)	medium (15-25°C)	high (more than 25°C)
Desert/arid	4 - 6	7 - 8	9 - 10
Semi-arid	4 - 5	6 - 7	8 - 9
Sub-humid	3 - 4	5 - 6	7 - 8
Humid	1 - 2	3 - 4	5 - 6

It is important to note that the irrigation schedules are based on the crop water needs in the peak period. It is further assumed Chat little or no rainfall occurs during the growing season. Some examples on the use of Table are given below:

1. Estimate the irrigation schedule for groundnuts grown on a deep, clayey soil, in a hoc and dry climate.

Firstly, the climatic class has to be identified: climate 3 (ETo = 8-9 mm/day) represents a hot climate. Table above shows that for climate 3 the interval for groundnuts grown on a clayey soil is 6 days and the net irrigation depth is 50 mm. This means that every 6 days the groundnuts should receive a net irrigation application of 50 mm.

2. Estimate the irrigation schedule for spinach grown on a loamy soil, in an area with an average temperature of 12° C during the growing season.

The average temperature is low: climate 1 (ETo = 4-5 mm/day). Table above shows, with climate 1, for spinach, grown on a loamy soil an interval of 4 days and a net irrigation depth of 20 mm.

3. Estimate the irrigation schedule of sorghum grown on a sandy soil, in an area with a temperature range of 15-25° C during the growing season.

The average temperature is medium: climate 2 (ETo = 6-7 mm/day). Table above shows, with climate 2 for sorghum grown on a sandy soil, an irrigation interval of 6 days and a net irrigation depth of 40 mm.

Adjusting the Irrigation Schedule

a. Adjustments for the non-peak periods

The irrigation schedule, which is obtained using table above, is valid for the peak period; in other words, for the mid-season stage of the crop.

During the early growth stages, when the plants are small, the crop water need is less than during the mid-season stage. Therefore, it may be possible to irrigate during the early stages of crop growth, with the same frequency as during the mid-season, but with smaller irrigation applications. It is risky to give the same irrigation application as during the mid-season, but less frequently; the young plants may suffer from water shortage as their roots are not able to take up water from the lower layers of the root zone.

Dry harvested crops or crops which are allowed to die before harvest (for example grain maize) need less water during the late season stage than during the mid-season stage (the peak period). During the late season stage, the roots of the crops are fully developed and therefore the same amount of water can be stored in the root zone as during the mid-season stage. It is thus possible to irrigate during the late season stage less frequently but with the same irrigation depth as during the peak period.

In summary, in order to save water, it may be feasible to irrigate, during the early stages of the crop development, with smaller irrigation applications than during the peak period. During the late season stage it may be feasible to irrigate less frequently, in particular if the crop is harvested dry.

When adjusting the irrigation schedule for the non-peak periods, it should always be kept in mind that the irrigation schedules must be simple, in particular in surface irrigation schemes where many farmers are involved. It will often be necessary to discuss with the farmers, before implementing the irrigation schedule, the various alternatives and come to an agreement which best satisfies all parties involved.

b. Adjustment for climates with considerable rainfall daring the growing season

The schedules obtained from table above are based on the assumption that little or no rainfall occurs during the growing season. If the contribution from the rainfall is considerable during the growing season, the schedules need to be adjusted: usually by making the interval longer. It may also be possible to reduce the net irrigation depth. It is difficult to estimate to which values the interval and the irrigation depth should be adjusted. It is therefore suggested to use the simple calculation method, instead of the estimation method, in the case of significant rainfall during the growing season. Alternatively it is possible to adjust the irrigation schedule to the actual rainfall.

c. Adjustment for local irrigation practices or irrigation method used

It may happen that the net irrigation depth obtained from table above is not suitable for the local conditions. It may not be possible, for example, to infiltrate 70 mm with the irrigation method used locally. Tests may have shown that it is only possible to infiltrate some 50 mm per application.

In such cases, both the net irrigation depth and the interval must be adjusted simultaneously. For example, suppose that maize is grown on a clayey soil in a moderately warm climate. According to table above, the Interval is 10 days and the net irrigation depth is 70 mm. This corresponds to an irrigation water need of 70/10 = 7 mm/day.

Instead of giving 70 mm every 10 days, it is also possible to give:

63 mm every 9 days,

56 mm every 8 days,

49 mm every 7 days,

42 mm every 6 days etc.

This means that in the above example an interval of seven days is chosen with a net application depth of 49 mm.

d. Adjustment for shallow soils

A soil which is shallow can only store a little water, even if the soil is clayey. For shallow soils - sandy, loamy or clayey - the column "shallow and/or sandy soil" of Table should be used.

e. Adjustment for salt-affected soils

In the case of irrigating salt-affected soils, special attention needs to be given to the determination of the irrigation schedule.

Application of the Simple Calculation Method

The simple calculation method to determine the irrigation schedule is based on the estimated depth (in mm) of the irrigation applications, and the calculated irrigation water need of the crop over the growing season.

Unlike the estimation method, the simple calculation method is based on calculated irrigation water needs. Thus, the influence of the climate, i.e. temperature and rainfall, is more accurately taken into account. The result of the simple calculation method will therefore be more accurate than the result of the estimation method.

The simple calculation method to determine the irrigation schedule involves the following steps that are explained in detail below:

Step 1:	Estimate the net and gross irrigation depth (d) in mm.
Step 2:	Calculate the irrigation water need (IN) in mm, over the total growing season.
Step 3:	Calculate the number of irrigation applications over the total growing season.
Step 4:	Calculate the irrigation interval in days.

Step 1: Estimate the net and gross irrigation depth (d) in mm

The net irrigation depth is best determined locally by checking how much water is given per irrigation application with the local irrigation method and practice. If no local data are easily available, table above can be used to estimate the net irrigation depth (d net), in mm. As can be seen from the table, the net irrigation depth is assumed to depend only on the root depth of the crop and on the soil type. It must be noted that the d net values in the table are approximate values only. Also the root depth is best determined locally. If no data are available, table below can be used which gives an indication of the root depth of the major field crops.

Approximate net irrigation depths, in mm

	Shallow rooting crops	Medium rooting crops	Deep rooting crops
Shallow and/or sandy soil	15	30	40
Loamy soil	20	40	60
Clayey soil	30	50	70

Approximate root depth of the major field crops

Shallow rooting crops (30-60 cm):	Crucifers (cabbage, cauliflower, etc.), celery, lettuce, onions, pineapple, potatoes, spinach, other vegetables except beets, carrots, cucumber.
Medium rooting crops (50-100 cm):	Bananas, beans, beets, carrots, clover, cacao, cucumber, groundnuts, palm trees, peas, pepper, sisal, soybeans, sugarbeet, sunflower, tobacco, tomatoes.
Deep rooting crops (90-150 cm):	Alfalfa, barley, citrus, cotton, dates, deciduous orchards, flax, grapes, maize, melons, oats, olives, safflower, sorghum, sugarcane, sweet potatoes, wheat.

Not all water, which is applied to the field, can indeed be used by the plants. Part of the water is lost through deep percolation and runoff. To reflect this water loss, the field application efficiency (ea) is used. The gross irrigation depth (d gross), in mm, takes into account the water loss during the irrigation application and is determined using the following formula:

$$d\,grass = \frac{100 \bullet d\,net}{ea}$$

d gross = gross irrigation depth in mm

d net = net irrigation depth in mm

ea = field application efficiency in percent

If reliable local data are available on the field application efficiency, these should be used. If such data are not available, the following values for the field application efficiency can be used:

- for surface irrigation	: ea = 60%
- for sprinkler irrigation	: ea = 75%
- for drip irrigation	: ea = 90%

If, for example, tomatoes are grown on a loamy soil, Tables 4 and 5 show that the estimated net irrigation depth is 40 mm. If furrow irrigation is used, the field application efficiency is 60% and the gross irrigation depth is determined as follows:

$$d\,grass = \frac{100 \bullet 40}{60} = 67\,mm = rounded\,65\,mm$$

Step:2 Calculate the irrigation water need (IN) in - over the total growing season

Assume that the irrigation water need (in mm/month) for tomatoes, planted 1 February and harvested 30 June, is as follows:

	Feb.	Mar.	Apr.	May	June
IN (mm/month)	IN (mm/month)	IN (mm/month)	IN (mm/month)	IN (mm/month)	IN (mm/month)

The irrigation water need of tomatoes for the total growing season (Feb-June) is thus (67 + 110 + 166 + 195 + 180 =) 718 mm. This means that over the total growing season a net water layer of 718 mm has to be brought onto the field.

If no data on irrigation water needs are available, the estimation method should be used.

Step 3: Calculate the number of irrigation applications over the total growing season

The number of irrigation applications over the total growing season can be obtained by dividing the irrigation water need over the growing season (Step 2) by the net irrigation depth per application (Step 1).

If the net depth of each irrigation application is 40 mm (d net = 40 mm; Step 1), and the irrigation water need over the growing season is 718 mm (Step 2), then a total of (718/40 =) 18 applications are required.

Step 4: Calculate the irrigation interval (INT) in days

Thus a total of 18 applications is required. The total growing season for tomatoes is 5 months (Feb-June) or 5 x 30 = 150 days. Eighteen applications in 150 days corresponds to one application every 150/18 = 8.3 days.

In other words, the interval between two irrigation applications is 8 days. To be on the safe side, the interval is always rounded off to the lower whole figure: for example 7.6 days becomes 7 days; 3.2 days becomes 3 days.

In this example, the irrigation schedule for tomatoes is as follows:

d net = 40 mm,

d gross = 65 mm,

interval = 8 days.

Adjusting the Simple Calculation Method for the Peak Period

When using the simple calculation method to determine the irrigation schedule, it is advisable to ensure that the crop does not suffer from undue water shortage in the months of peak irrigation water need.

For instance, in the above example the interval is 8 days, while the net irrigation depth is 40 mm. Thus every 30 days (or each month): 30/8 x 40 mm = 150 mm water is applied. The amount of water given during each month (d net) should be compared with the amount of irrigation water needed during that month (IN).

The result is shown below. The "IN" values represents the irrigation water needs, while the "d net" values represent the amount of water applied. The "d net - IN" values show whether too much or too little water has been applied:

	Feb	Mar	Apr	May	June	Total
IN (mm/month)	87	110	166	195	180	718
d net (mm/month)	150	150	150	150	150	750
d net - IN (mm/month)	+83	+40	-16	-45	-30	+32

The total net amount of irrigation water applied (750 mm) is more than sufficient to cover the total irrigation water need (718 mm). However, in February and March too much water has been applied, while in April, May and June, too little water has been applied.

Care should be taken with under-irrigation (too little irrigation) in the peak period as this period normally coincides with the growth stages of the crops that are most sensitive to water shortages.

To overcome the risk of water shortages in the peak months, it is possible to refine the simple calculation method by looking only at the months of peak irrigation water need and basing the determination of the interval on the peak period only.

In the example given above for tomatoes, this means looking at the months April, May and June:

Months of peak irrigation water need	Apr	May	June	Sub-total
IN (mm/month)	166	195	180	541

The total irrigation water need from April to June (90 days) is 541 mm, while the net irrigation depth is 40 mm. Thus 541/40 = 13.5 (rounded 14) applications are needed. Fourteen applications in 90 days means one application every 6.4 (rounded 6) days. Calculated this way the irrigation schedule for the tomatoes would be:

d net = 40 mm,

d gross = 65 mm,

interval = 6 days.

Over the total growing period of 150 days, this means 150/6 = 25 applications, each 40 mm net and thus in total 25 x 40 = 1000 mm.

The overall result of adjusting the irrigation schedule to the months of peak irrigation water demand is shown below:

	Feb	Mar	Apr	May	June	Total
IN (mm/month)	67	110	166	195	180	718
d net (mm/month)	200	200	200	200	200	1000
d net - IN (mm/month)	+133	+90	+34	+5	+20	+282

This way of determining the irrigation schedule avoids water shortages in the month of peak water needs but on the other hand also results in a higher seasonal irrigation water application.

It is possible to combine the two schedules. In this way some water is saved, and there are no water shortages in the peak period, but it is a bit more complicated for the farmers.

The result of the combined irrigation schedule for the whole growing season is as follows:

	Feb	Mar	Apr	May	June	Total
IN (mm/month)	67	110	166	195	180	718
d net (mm/month)	150	150	200	200	200	900
d net - IN (mm/month)	+83	+90	+40	+5	+20	+182

In summary:

Feb-March

d net = 40 mm,

d gross = 65 mm,

Interval = 8 days.

April-May-June

d net = 40 mm,

d gross = 65 mm,

Interval = 6 days.

Adjusting the Irrigation Schedule to Actual Rainfall

The estimation method to determine the irrigation schedule can only be used when no significant rainfall occurs during the growing season. The simple calculation method is based on the average irrigation water need of the crop which is the average crop water need minus the average effective rainfall. This method is used when designing and implementing an irrigation system with a "rotational" water supply: each field receives a certain amount of water on dates that are already fixed in advance. The rotational supply takes into account the average rainfall only and thus does not take into account the actual rainfall; this results in over-irrigation in wetter than average years and under-irrigation in drier than average years. In surface irrigation systems the rotational water supply method is most commonly used.

There are also water supply methods which allow the irrigation water to be distributed "on demand". The farmer can take water whenever necessary. In this case it is possible to take the actual rainfall into account and thus give the correct amount of irrigation water even in drier or wetter years. With this method of irrigation scheduling, however, the rainfall has to be measured on a daily basis. The net irrigation depth (d net) has to be determined in accordance with the irrigation method used. In addition, the crop water need has to be known on a daily basis for each month of the growing season. As soon as the accumulated water deficit exceeds the value of the net irrigation depth, irrigation water is supplied.

An example is given below for a situation with a crop water need (CWN) of 8 am/day and a net irrigation depth (d net) of 45 mm. As soon as the accumulated deficit exceeds the d net (= 45 mm), irrigation water is supplied. Note that the "deficit" can never be positive; maximum zero.

day	CWN (mm/day)	Rain (mm)	d net (mm)		Accumulated deficit (mm)
1	8	-	-		-8
2	8	-	-	(-8-8)	-16
3	8	-	-	(-16-8)	-24
4	8	-	-	(-24-8)	-32
5	8	-	-	(-32-8)	-40
6	8	-	45	(-40-8+45)	-3
7	8	-	-	(-3-8)	-11

8	8	12	-	(-11-8+12)	-7
9	8	24	-	(-7-8+24)	0
10	8	-	-	(0-8)	-8
11	8	-	-	(-8-8)	-16
12	8	-	-	(-16-8)	-24
13	8	4	-	(-24-8+4)	-28
14	8	-	-	(-28-8)	-36
15	8	-	-	(-36-8)	-44
16	8	-	45	(-44-8+45)	-7
17	8	-	-	(-7-8)	-15
etc.					

In the above example of adjusting the irrigation schedule to the actual rainfall, irrigation takes place on day 6, on day 16, etc. with on each occasion a net irrigation depth of 45 mm.

Advantages of Irrigation Scheduling

1. It enables the farmer to schedule water rotation among the various fields to\ minimize crop water stress and maximize yields.
2. It reduces the farmer's cost of water and labor through fewer irrigations, thereby making maximum use of soil moisture storage.
3. It lowers fertilizer costs by holding surface runoff and deep percolation (leaching) to a minimum.
4. It increases net returns by increasing crop yields and crop quality.
5. It minimizes water-logging problems by reducing the drainage requirements.
6. It assists in controlling root zone salinity problems through controlled leaching.
7. It results in additional returns by using the "saved" water to irrigate non-cash crops that otherwise would not be irrigated during water-short periods.

Ditch Irrigation

Irrigation ditches are manmade channels that deliver water to homes, farms, industries and other human uses. Most ditches divert water from natural creeks and rivers and bring it to other areas.

Ditches have headgates on the creeks that they divert. Most headgates are operated manually by ditch company personnel. Ditch headgates can be closed, when needed, to prevent water from being diverted from the creek into the ditch.

Ditches are typically owned, operated, and maintained by private companies. Water in ditches is allocated by shares issued by the company. Those shares represent proportional amounts of water.

Only those who have rights or shares in the ditch are allowed to remove water from it.

Ditch companies are responsible for keeping the ditch channels clear of debris or sediment that would obstruct the flow of water. Barrow (roadside) ditches are the maintenance responsibility of the adjacent property owner – including culverts that run under driveways.

Overhead Irrigation

Overhead sprinkler systems spray water into the air above and around the foliage of the crop in a broadcast pattern. The water droplets should be large enough to fall through the canopy to growing substrate or to run along the foliage and stems to the base of the crop.

Sizes of droplets are largely determined by the sprinkler system to achieve distribution. Overhead sprinklers are 70 to 75 percent efficient in water delivery; evaporation occurs in the air, from the plant foliage, and from the ground surface.

The circular pattern of application makes it difficult to achieve a high coefficient of uniformity in delivering water to all plants. Sprinkler nozzles must be placed with sufficient overlap (between 40 to 60 percent) to achieve nearly uniform coverage of the surface area. More overlap is needed for windy situations.

Sprinklers usually throw water beyond the structure. Over-irrigation often occurs (i.e. the duration of the irrigation event is too long) since the containers on the fringes have to be adequately wet.

Many overwintering structures have a single row of sprinklers down the center. These sprinklers are much closer together in order to wet the interior area of the structure more uniformly. Containers are spread out under the part of the broadcast covered area where the water coverage is most uniform.

Sprinkler nozzles are designed to function at some pressure and discharge rate which determines the diameter of water throw. Operating pressure may range from 20 psi to 80 psi for large sprinklers. Discharge rates may range from 0.5 gallons per minute (gpm) to 1,000 gpm for a large gun. Wetting diameters range from 35 feet to 300 feet.

The application rate in inches per hour (iph) onto the ground is important. A rate of 0.2 iph to 0.5 iph is common. The crop should be watered fairly quickly to reduce the evaporative losses to the air and to reduce the time the foliage is wetted. At the same time, the irrigation should not cause runoff or puddles in fields. In container production an impervious surface may mean there will be runoff but it should be controlled.

It is a challenge to try to match sprinklers to any small or narrow area. In an overwintering (nursery) house, the sprinkler heads are placed close together in a line to get multiple overlaps of heads over some width of the structure. The goal is to achieve uniform coverage over the width of the structure. The many overlaps increase the application rate of the water and substrate wetting may suffer. Water applied too quickly may drain through quickly. Water will be thrown beyond the structure but the application rate will be low at the outer edge of the sprinkler pattern.

Overhead sprinklers are efficient for bedding plants, field crops, and small container plants placed close together. However, foliage is wet during each irrigation, so disease pressure may be a problem. Irrigations should be shceduled in time for the foliage to dry before relative humididty rises at night (in humid environments). In larger fields, the sprinklers can be placed in many rows so that overlapping of nozzles is possible and application uniformity can be good.

Overhead sprinklers include fixed installations, center pivots, traveling guns, and portable sprinklers. Each system has its own characteristics to consider.

Agricultural Water Management

Agricultural water management includes the management of water used in crop production (both rainfed and irrigated), livestock production and inland fisheries. Improved agricultural water management in these production areas is the answer to both global food security and poverty reduction.

The main challenge confronting water management in agriculture is to improve water use efficiency and its sustainability.

This can be achieved through (i) an increase in crop water productivity (an increased in marketable crop yield per unit of water transpired) through irrigation, (ii) a decrease in water losses through soil evaporation that could otherwise be used by plants for their growth, and (iii) an increase in soil water storage within the plant rooting zone through better soil and water management practices at farm and area-wide (catchment) scales.

Tracking and quantifying water fluxes at different spatial and temporal scales within the plant rooting zone remains a formidable challenge because of the interactions between water sources from rainfall, irrigation and subsurface water on plant uptake, soil evaporation, plant transpiration (water transpired by plants) and runoff or drainage losses from crop-growing areas. The use of isotopic and nuclear techniques to investigate the relative importance of soil and irrigation management factors that influence these interactions will greatly assist in the development of water management packages that involve the consideration of soil nutrient status, type of crops grown, growth stages and the overall agro-ecosystems to minimize not only water but also nutrient losses from the farmlands and enhance water and nutrient use efficiencies in agro-ecosystems under both rainfed and irrigated conditions.

Many nuclear and isotopic techniques are being employed in soil water management studies. The soil moisture neutron probe is ideal in field-scale rooting zone measurement of soil water, providing accurate data on the availability of water for determining crop water use and water use efficiency and for establishing optimal irrigation scheduling under different cropping systems especially under saline conditions.

The use of oxygen-18, hydrogen-2 (deuterium) and other isotopes is an integral part of agricultural water management, allowing the identification of water (and plant nutrients) sources and the tracking of water movement and pathways within agricultural landscapes as influenced by different irrigation technologies, cropping systems and farming practices. It also helps in the understanding of plant water use, quantifying crop transpiration and soil evaporation and allows us to devise strategies to improve crop production, reduce unproductive water losses and prevent land and water degradation.

References

- Safe-uses-of-agricultural-water: extension.psu.edu, Retrieved 16, May 2020
- Water-food-agriculture: lenntech.com, Retrieved 25, June 2020
- Topic-water-management: naweb.iaea.org, Retrieved 23, May 2020
- Irrigation-ditches-fact-sheet-1-201404141426: static.bouldercolorado.gov, Retrieved 09, March 2020
- Irrigation-scheduling, potato: cropwatch.unl.edu, Retrieved 27, April 2020

Chapter 5

Machinery and Technologies used in Agriculture

All the mechanical structures and devices used in agriculture fall under agricultural machinery. Different types of machines used in farming are manure spreader, cultivator, plough, planter, threshing machine, etc. This chapter has been carefully written to provide an easy understanding of the varied facets of these machines and technologies used in agriculture.

Agricultural Machinery

Agricultural machinery is machinery used in farming or other agriculture. There are many types of such equipment, from hand tools and power tools to tractors and the countless kinds of farm implements that they tow or operate. Diverse arrays of equipment are used in both organic and nonorganic farming. Especially since the advent of mechanised agriculture, agricultural machinery is an indispensable part of how the world is fed.

A German combine harvester by Claas.

History of the Machines

The Industrial Revolution

With the coming of the Industrial Revolution and the development of more complicated machines, farming methods took a great leap forward. Instead of harvesting grain by hand with a sharp blade, wheeled machines cut a continuous swath. Instead of threshing the grain by beating it with sticks, threshing machines separated the seeds from the heads and stalks. The first tractors appeared in the late 19th century.

Steam Power

Power for agricultural machinery was originally supplied by ox or other domesticated animals. With the invention of steam power came the portable engine, and later the traction engine, a multipurpose, mobile energy source that was the ground-crawling cousin to the steam locomotive. Agricultural steam engines took over the heavy pulling work of oxen, and were also equipped with a pulley that could power stationary machines via the use of a long belt. The steam-powered machines were low-powered by today's standards but, because of their size and their low gear ratios, they could provide a large drawbar pull. Their slow speed led farmers to comment that tractors had two speeds: "slow, and damn slow."

Internal Combustion Engines

The internal combustion engine; first the petrol engine, and later diesel engines; became the main source of power for the next generation of tractors. These engines also contributed to the development of the self-propelled, combined harvester and thresher, or combine harvester (also shortened to 'combine'). Instead of cutting the grain stalks and transporting them to a stationary threshing machine, these combines cut, threshed, and separated the grain while moving continuously through the field.

Types

A John Deere cotton harvester at work in a cotton field.

Combines might have taken the harvesting job away from tractors, but tractors still do the majority of work on a modern farm. They are used to push implements—machines that till the ground, plant seed, and perform other tasks.

Tillage implements prepare the soil for planting by loosening the soil and killing weeds or competing plants. The best-known is the plow, the ancient implement that was upgraded in 1838 by John Deere. Plows are now used less frequently in the U.S. than formerly, with offset disks used instead to turn over the soil, and chisels used to gain the depth needed to retain moisture.

The most common type of seeder is called a planter, and spaces seeds out equally in long rows, which are usually two to three feet apart. Some crops are planted by drills, which put out much more seed in rows less than a foot apart, blanketing the field with crops. Transplanters automate the task of transplanting seedlings to the field. With the widespread use of plastic mulch, plastic mulch layers, transplanters, and seeders lay down long rows of plastic, and plant through them automatically.

From left to right: John Deere 7800 tractor with Houle slurry trailer, Case IH combine harvester, New Holland FX 25 forage harvester with corn head.

After planting, other implements can be used to cultivate weeds from between rows, or to spread fertilizer and pesticides. Hay balers can be used to tightly package grass or alfalfa into a storable form for the winter months.

Modern irrigation relies on machinery. Engines, pumps and other specialized gear provide water quickly and in high volumes to large areas of land. Similar types of equipment can be used to deliver fertilizers and pesticides.

Besides the tractor, other vehicles have been adapted for use in farming, including trucks, airplanes, and helicopters, such as for transporting crops and making equipment mobile, to aerial spraying and livestock herd management.

New Technology and the Future

The basic technology of agricultural machines has changed little in the last century. Though modern harvesters and planters may do a better job or be slightly tweaked from their predecessors, the US$250,000 combine of today still cuts, threshes, and separates grain in the same way it has always been done. However, technology is changing the way that humans operate the machines, as computer monitoring systems, GPS locators, and self-steer programs allow the most advanced tractors and implements to be more precise and less wasteful in the use of fuel, seed, or fertilizer. In the foreseeable future, there may be mass production of driverless tractors, which use GPS maps and electronic sensors.

Open Source Agricultural Equipment

Many farmers are upset by their inability to fix the new types of high-tech farm equipment. This is due mostly to companies using intellectual property law to prevent farmers from having the legal right to fix their equipment (or gain access to the information to allow them to do it). This has encouraged groups such as Open Source Ecology and Farm Hack to begin to make open source agricultural machinery. In addition on a smaller scale FarmBot and the RepRap open source 3D printer community has begun to make open-source farm tools available of increasing levels of sophistication. In October 2015 an exemption was added to the DMCA to allow inspection and modification of the software in cars and other vehicles including agricultural machinery.

Mechanised Agriculture

A cotton picker at work. The first successful models were introduced in the mid-1940s and each could do the work of 50 hand pickers.

Mechanised agriculture is the process of using agricultural machinery to mechanise the work of agriculture, greatly increasing farm worker productivity. In modern times, powered machinery has replaced many farm jobs formerly carried out by manual labour or by working animals such as oxen, horses and mules.

The entire history of agriculture contains many examples of the use of tools, such as the hoe and the plough. But the ongoing integration of machines since the Industrial Revolution has allowed farming to become much less labour-intensive.

Current mechanised agriculture includes the use of tractors, trucks, combine harvesters, countless types of farm implements, aeroplanes and helicopters (for aerial application), and other vehicles. Precision agriculture even uses computers in conjunction with satellite imagery and satellite navigation (GPS guidance) to increase yields.

Mechanisation was one of the large factors responsible for urbanisation and industrial economies. Besides improving production efficiency, mechanisation encourages large scale production and sometimes can improve the quality of farm produce. On the other hand, it can displace unskilled farm labour and can cause environmental degradation (such as pollution, deforestation, and soil erosion), especially if it is applied shortsightedly rather than holistically.

History

A reaper at Woolbrook, New South Wales.

Threshing machine in 1881. Steam engines were also used to power threshing machines. Today both reaping and threshing are done with a combine harvester.

Jethro Tull's seed drill (ca. 1701) was a mechanical seed spacing and depth placing device that increased crop yields and saved seed. It was an important factor in the British Agricultural Revolution.

Since the beginning of agriculture threshing was done by hand with a flail, requiring a great deal of labour. The threshing machine, which was invented in 1794 but not widely used for several more decades, simplified the operation and allowed the use of animal power. Before the invention of the grain cradle (ca. 1790) an able bodied labourer could reap about one quarter acre of wheat in a day using a sickle. It was estimated that for each of Cyrus McCormick's horse pulled reapers (ca. 1830s) freed up five men for military service in the US Civil War. Later innovations included raking and binding machines. By 1890 two men and two horses could cut, rake and bind 20 acres of wheat per day.

In the 1880s the reaper and threshing machine were combined into the combine harvester. These machines required large teams of horses or mules to pull. Steam power was applied to threshing machines in the late 19th century. There were steam engines that moved around on wheels under their own power for supplying temporary power to stationary threshing machines. These were called *road engines,* and Henry Ford seeing one as a boy was inspired to build an automobile.

With internal combustion came the first modern tractors in the early 1900s, becoming more popular after the Fordson tractor (ca. 1917). At first reapers and combine harvesters were pulled by tractors, but in the 1930s self powered combines were developed.

Advertising for motorised equipment in farm journals during this era did its best to compete against horse-drawn methods with economic arguments, extolling common themes such as that a tractor "eats only when it works", that one tractor could replace many horses, and that mechanisation could allow one man to get more work done per day than he ever had before. The horse population in the US began to decline in the 1920s after the conversion of agriculture and transportation to internal combustion. Peak tractor sales in the US were around 1950. In addition to saving labour, this freed up much land previously used for supporting draft animals. The greatest period of growth in agricultural productivity in the US was from the 1940s to the 1970s, during which time agriculture was benefiting from internal combustion powered tractors and combine harvesters, chemical fertilizers and the green revolution.

Although farmers of corn, wheat, soy, and other commodity crops had replaced most of their workers with harvesting machines and combines enabling them to efficiently cut and gather grains, growers of produce continued to rely on human pickers to avoid the bruising of the product in order to maintain the blemish-free appearance demanded of consumers. The continuous supply of illegal workers from Latin America that were willing to harvest the crops for low wages further suppressed the need for mechanisation. As the number of illegal workers has continued to decline since reaching its peak in 2007 due to increased border patrols and an improving Mexican economy, the industry is increasing the use of mechanisation. Proponents argue that mechanisation will boost productivity and help to maintain low food prices while farm worker advocates assert that it will eliminate jobs and will give an advantage to large growers who are able to afford the required equipment.

Current Status of Future Applications

Asparagus Harvesting

Asparagus are presently harvested by hand with labour costs at 71% of production costs and 44% of selling costs. Asparagus is a difficult crop to harvest since each spear matures at a different speed making it difficult to achieve a uniform harvest. A prototype asparagus harvesting machine - using a light-beam sensor to identify the taller spears - is expected to be available for commercial use.

Blueberry Harvesting

Mechanization of Maine's blueberry industry has reduced the number of migrant workers required from 5,000 in 2005 to 1,500 in 2015 even though production has increased from 50-60 million pounds per year in 2005 to 90 million pounds in 2015.

Chili Pepper Harvesting

As of 2014, prototype chili pepper harvesters are being tested by New Mexico State University. The New Mexico green chile crop is currently hand-picked entirely by field workers as chili pods tend to bruise easily. The first commercial application commenced in 2015. The equipment is expected to increase yield per acre and help to offset a sharp decline in acreage planted due to the lack of available labour and drought conditions.

Orange Harvesting

As of 2010, approximately 10% of the processing orange acreage in Florida is harvested mechanically. Mechanisation has progressed slowly due to the uncertainty of future economic benefits due to competition from Brazil and the transitory damage to orange trees when they are harvested.

Raisin Harvesting

As of 2007, mechanised harvesting of raisins is at 45%; however the rate has slowed due to high raisin demand and prices making the conversion away from hand labour less urgent. A new strain of grape developed by the USDA that drys on the vine and is easily harvested mechanically is expected to reduce the demand for labour.

Strawberry Harvesting

Strawberries are a high cost-high value crop with the economics supporting mechanisation. In 2005, picking and hauling costs were estimated at $594 per ton or 51% of the total grower cost. However, the delicate nature of fruit make it an unlikely candidate for mechanisation in the near future. A strawberry harvester developed by Shibuya Seiki and unveiled in Japan in 2013 is able to pick a strawberry every eight seconds. The robot identifies which strawberries are ready to pick by using three separate cameras and then once identified as ready, a mechanised arm snips the fruit free and gently places it in a basket. The robot moves on rails between the rows of strawberries which are generally contained within elevated greenhouses. The machine costs 5 million yen. A new strawberry harvester made by Agrobot that will harvest strawberries on raised, hydroponic beds using 60 robotic arms is expected to be released in 2016.

Tomato Harvesting

Mechanical harvesting of tomatoes started in 1965 and as of 2010, nearly all processing tomatoes are mechanically harvested. As of 2010, 95% of the US processed tomato crop is produced in California. Although fresh market tomatoes have substantial hand harvesting costs (in 2007, the costs of hand picking and hauling were $86 per ton which is 19% of total grower cost), packing and selling costs were more of a concern (at 44% of total grower cost) making it likely that cost saving efforts would be applied there.

According to a 1977 report by the California Agrarian Action Project, during the summer of 1976 in California, many harvest machines had been equipped with a photo-electric scanner that sorted out green tomatoes among the ripe red ones using infrared lights and colour sensors. It worked in lieu of 5,000 hand harvesters causing displacement of innumerable farm labourers as well as wage cuts and shorter work periods. Migrant workers were hit the hardest. To withstand the rigour of the machines, new crop varieties were bred to match the automated pickers. UC Davis Professor G.C. Hanna propagated a thick-skinned tomato called VF-145. But even still, millions were damaged with impact cracks and university breeders produced a more tougher and juiceless "square round" tomato. Small farms were of insufficient size to obtain financing to purchase the equipment and within 10 years, 85% of the state's 4,000 cannery tomato farmers were out of the business. This led to a concentrated tomato industry in California that "now packed 85% of the nation's tomato products". The monoculture fields fostered rapid pest growth, requiring the use of "more than four million pounds of pesticides each year" which greatly affected the health of the soil, the farm workers, and possibly the consumers.

Manure Spreader

A modern manure spreader.

A manure spreader or muck spreader or honey wagon is an agricultural machine used to distribute manure over a field as a fertilizer. A typical (modern) manure spreader consists of a trailer towed behind a tractor with a rotating mechanism driven by the tractor's power take off (PTO). Truck mounted manure spreaders are also common in North America.

Operation

Manure spreaders began as ground-driven units which could be pulled by a horse or team of horses. Many of these ground-driven spreaders are still produced today, mostly in the form of small units that can be pulled behind a larger garden tractor or an all terrain vehicle (ATV). In recent years hydraulic and PTO driven units have been developed to offer variable application rates. Several models are also designed with removable rotating mechanisms (beaters), attachable side extensions, and tailgates for hauling chopped forages, cereal grains, and other crops. A typical (modern) manure spreader consists of a trailer towed behind a tractor with a rotating mechanism driven by the tractor's power take off (PTO).

History

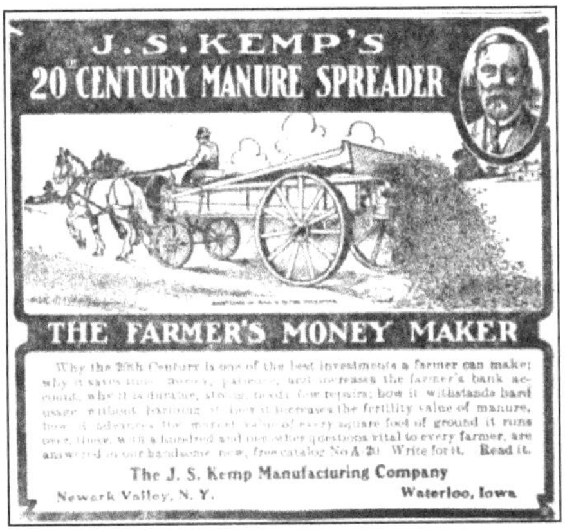

An advertisement for a J.S.Kemp model spreader.

The first successful automated manure spreader was designed by Joseph Kemp in 1875. Manure spreaders began as ground-driven units which could be pulled by a horse or team of horses. At the time of his invention, he was living in Waterloo, Canada, but thereafter, he moved to Newark Valley, NY and formed the J.S. Kemp Manufacturing Co. to manufacture and market his current and subsequent designs. In 1903, he expanded the company to Waterloo, Iowa before selling the design to International Harvester, in 1906.

The original New Idea spreader design. Note the paddle system at the rear that creates the 'widespreading' effect.

Joseph Oppenheim of Maria Stein, Ohio was the inventor of the first modern 'widespreading' manure spreader and is honored as such in the Ohio Agricultural Hall of Fame. Originally manure was thrown from a wagon. Later, "manure unloaders" used a drag chain at the bottom of the wagon to pull the load of manure to the rear where it was shredded by a pair of beaters. Because the unloaders deposited manure directly behind the wagon but with very little spreading to the sides, farmers still had to take the time-consuming step of heading into the fields with peg-tooth drags or similar implements to spread the manure in order to prevent burning the soil.

Oppenheim, a schoolmaster in the small town, concerned that his older male students often missed school loading and spreading manure, patented a wagon that, behind the drag chain and two beaters, incorporated a steel axle with several wooden paddles attached to the shaft at an angle to throw the manure outward in a broad pattern eliminating the necessity for manual spreading. On October 18, 1899, Oppenheim began to produce his new manure spreader, incorporating the "widespread" paddle device. Neighbors soon referred to it as "Oppenheim's new idea" and Oppenheim adopted this name for his business.

Although Oppenheim died in November, 1901, the demand for the New Idea Spreader Company's labor-saving "widespread" machines quickly grew and fifteen years later, under the direction of his oldest son, B.C. Oppenheim, and Henry Synck, one of Oppenheim's first employees, the company, had branches in eight states and an assembly plant in Guelph, Ontario. It had total sales in 1916 of $1,250,000. Eight years later, in 1924, the factory was turning out 125 manure spreaders in an eight-hour day. and "became the brand that set the standards for spreader performance, durability and reliability decade after decade."

During the 1920s, Henry Synck, who became president of the company in 1936 when Joseph's son, B.C. Oppenheim, died, patented several improvements to the spreader. In 1945 the Oppenheim family sold its controlling interest in the closely held New Idea Company to AVCO Manufacturing. AVCO later sold the company to White Farm Equipment Company which in 1993 sold it to AGCO (Allis-Gleaner Corporation), the current owner.

It is clear, however, that there were other competitors in this field, each of whom spread the manure by a slightly different technique. One of these is the Great Western Farm Equipment Line, produced in Chicago, IL.

In Popular Culture

In the post-Civil War AMC TV series *Hell on Wheels*, the muck wagon is a vehicle used to collect dead bodies, animal carcasses, and filled chamber pots. When The Swede is disgraced, Durant demotes him from head of security to driver of the muck wagon.

Cultivator

A cultivator is any of several types of farm implement used for secondary tillage. One sense of the name refers to frames with the teeth (also called shanks) that pierce the soil as they are dragged through it linearly. Another sense refers to machines that use rotary motion of disks or teeth to accomplish a similar result. The rotary tiller is a principle example.

A cultivator pulled by a tractor in Canada in 1943.

Cultivators stir and pulverize the soil, either before planting (to aerate the soil and prepare a smooth, loose seedbed) or after the crop has begun growing (to kill weeds—controlled disturbance of the topsoil close to the crop plants kills the surrounding weeds by uprooting them, burying their leaves to disrupt their photosynthesis, or a combination of both). Unlike a harrow, which disturbs the entire surface of the soil, cultivators are designed to disturb the soil in careful patterns, sparing the crop plants but disrupting the weeds.

Cultivators of the toothed type are often similar in form to chisel plows, but their goals are different. Cultivator teeth work near the surface, usually for weed control, whereas chisel plow shanks work deep beneath the surface, breaking up hardpan. Consequently, cultivating also takes much less power per shank than does chisel plowing.

Small toothed cultivators pushed or pulled by a single person are used as garden tools for small-scale gardening, such as for the household's own use or for small market gardens. Similarly sized rotary tillers combine the functions of harrow and cultivator into one multipurpose machine.

Cultivators are usually either self-propelled or drawn as an attachment behind either a two-wheel tractor or four-wheel tractor. For two-wheel tractors they are usually rigidly fixed and powered via couplings to the tractors' transmission. For four-wheel tractors they are usually attached by means of a three-point hitch and driven by a power take-off (PTO). Drawbar hookup is also still commonly used worldwide. Draft-animal power is sometimes still used today, being somewhat common in developing nations although rare in more industrialized economies.

History

The basic idea of soil scratching for weed control is ancient and was done with hoes or mattocks for millennia before cultivators were developed. Cultivators were originally drawn by draft animals (such as horses, mules, or oxen) or were pushed or drawn by people. In modern commercial agriculture, the amount of cultivating done for weed control has been greatly reduced via use of herbicides instead. However, herbicides are not always desirable—for example, in organic farming.

The powered rotary hoe was invented by Arthur Clifford Howard who, in 1912, began experimenting with rotary tillage on his father's farm at Gilgandra, New South Wales, Australia. Initially using his father's steam tractor engine as a power source, he found that ground could be mechanically tilled without soil-packing occurring, as was the case with normal ploughing. His earliest designs threw the tilled soil sideways, until he improved his invention by designing an L-shaped blade mounted on widely spaced flanges fixed to a small-diameter rotor. With fellow apprentice Everard McCleary, he established a company to make his machine, but plans were interrupted by World War I. In 1919 Howard returned to Australia and resumed his design work, patenting a design with 5 rotary hoe cultivator blades and an internal combustion engine in 1920.

In March 1922, Howard formed the company Austral Auto Cultivators Pty Ltd, which later became known as Howard Auto Cultivators. It was based in Northmead, a suburb of Sydney, from 1927.

Meanwhile, in North America during the 1910s, tractors were evolving away from traction engine-sized monsters toward smaller, lighter, more affordable machines. The Fordson tractor especially had made tractors affordable and practical for small and medium family farms for the first time in history. Cultivating was somewhat of an afterthought in the Fordson's design, which reflected the fact that even just bringing practical motorized tractive power alone to this market segment was in itself a milestone. This left an opportunity for others to pursue better motorized cultivating. Between 1915 and 1920, various inventors and farm implement companies experimented with a class of machines referred to as *motor cultivators*, which were simply modified horse-drawn shank-type cultivators with motors added for self-propulsion. This class of machines found limited market success. But by 1921 International Harvester had combined motorized cultivating with the other tasks of tractors (tractive power and belt work) to create the Farmall, the general-purpose tractor tailored to cultivating that basically invented the category of row-crop tractors.

In Australia, by the 1930s, Howard was finding it increasingly difficult to meet a growing worldwide demand for exports of his machines. He travelled to the United Kingdom, founding the company Rotary Hoes Ltd in East Horndon, Essex, in July 1938. Branches of this new company subsequently opened in the United States of America, South Africa, Germany, France, Italy, Spain, Brazil, Malaysia, Australia and New Zealand. It later became the holding company for Howard Rotavator Co. Ltd. The Howard Group of companies was acquired by the Danish Thrige Agro Group in 1985, and in December 2000 the Howard Group became a member of Kongskilde Industries of Soroe, Denmark.

When herbicidal weed control was first widely commercialized in the 1950s and 1960s, it played into that era's optimistic worldview in which sciences such as chemistry would usher in a new age of modernity that would leave old-fashioned practices (such as weed control via cultivators) in the dustbin of history. Thus herbicidal weed control was adopted very widely, and in some cases too heavily and hastily. In subsequent decades, people overcame this initial imbalance and came to realize that herbicidal weed control has limitations and externalities, and it must be managed intelligently. It is still widely used, and probably will continue to be indispensable to affordable food production worldwide for the foreseeable future; but its wise management includes seeking alternate methods, such as the traditional standby of mechanical cultivation, where practical.

Industrial Use

To the extent that cultivating is done commercially today (such as in truck farming), it is usually powered by tractors, especially row-crop tractors. Industrial cultivators can vary greatly in size and shape, from 10 feet (3 m) to 80 feet (24 m) wide. Many are equipped with hydraulic wings that fold up to make road travel easier and safer. Different types are used for preparation of fields before planting, and for the control of weeds between row crops. The cultivator may be an implement trailed after the tractor via a drawbar; mounted on the three-point hitch; or mounted on a frame beneath the tractor. Active cultivator implements are driven by a power take-off shaft. While most cultivator are considered a secondary tillage implement, active cultivators are commonly used for primary tillage in lighter soils instead of plowing. The largest versions available are about 6 m (20 ft) wide, and require a tractor with an excess of 150 horsepower (110 kW) (PTO) to drive them.

Field cultivators are used to complete tillage operations in many types of arable crop fields. The main function of the field cultivator is to prepare a proper seedbed for the crop to be planted into, to bury crop residue in the soil (helping to warm the soil before planting), to control weeds, and to mix and incorporate the soil to ensure the growing crop has enough water and nutrients to grow well during the growing season. The implement has many shanks mounted on the underside of a metal frame, and small narrow rods at the rear of the machine that smooth out the soil surface for easier travel later when planting. In most field cultivators, one-to-many hydraulic cylinders raise and lower the implement and control its depth.

Row Crop Cultivators

Home made sweep. Notice the inner and outer “sweep” blades.

The main function of the row crop cultivator is weed control between the rows of an established crop. Row crop cultivators are usually raised and lowered by a three-point hitch and the depth is controlled by gauge wheels.

Sometimes referred to as *sweep cultivators*, these commonly have two center blades that cut weeds from the roots near the base of the crop and turn over soil, while two rear sweeps further outward than the center blades deal with the center of the row, and can be anywhere from 1 to 36 rows wide.

Garden Cultivators

Small tilling equipment, used in small gardens such as household gardens and small commercial gardens, can provide both primary and secondary tillage. For example, a rotary tiller does both the "plowing" and the "harrowing", preparing a smooth, loose seedbed. It does not provide the row-wise weed control that cultivator teeth would. For that task, there are single-person-pushable toothed cultivators.

Variants and Trademarks

A Japanese two-wheel tractor.

Rotary tillers are a type of cultivators. Rotary tillers are popular with home gardeners who want large vegetable gardens. The garden may be tilled a few times before planting each crop. Rotary tillers may be rented from tool rental centers for single-use applications, such as when planting grass.

A small rotary hoe for domestic gardens was known by the trademark Rototiller and another, made by the Howard Group, who produced a range of rotary tillers, was known as the Rotavator.

Rototiller

> The small rototiller is typically propelled forward via a (1–5 horsepower or 0.8–3.5 kilowatts) petrol engine rotating the tines, and do not have powered wheels, though they may have small transport/level control wheel(s). To keep the machine from moving forward too fast, an adjustable tine is usually fixed just behind the blades so that through friction with deeper un-tilled soil, it acts

as a brake, slowing the machine and allowing it to pulverize the soils. The slower a rototiller moves forward, the more soil tilth can be obtained. The operator can control the amount of friction/braking action by raising and lowering the handlebars of the tiller. Rototillers do not have a reverse as such backwards movement towards the operator could cause serious injury. While operating, the rototiller can be pulled backwards to go over areas that were not pulverized enough, but care must be taken to ensure that the operator does not stumble and pull the rototiller on top of himself. Rototilling is much faster than manual tilling, but notoriously difficult to handle and exhausting work, especially in the heavier and higher horsepower models. If the rototiller's blades catch on unseen subsurface objects, such as tree roots and buried garbage, it can cause the rototiller to abruptly and violently move in any direction.

Rotavator

Unlike the Rototiller, the self-propelled Howard Rotavator is equipped with a gearbox and driven forward, or held back, by its wheels. The gearbox enables the forward speed to be adjusted while the rotational speed of the tines remains constant which enables the operator to easily regulate the extent to which soil is engaged. For a two-wheel tractor rotavator this greatly reduces the workload of the operator as compared to a rototiller. These rotavators are generally more heavy duty, come in higher power (4–18 horsepower or 3–13 kilowatts) with either petrol or diesel engines and can cover much more area per hour. The trademarked word "Rotavator" is one of the longest single-word palindromes in the English language.

Mini tiller

Mini tillers are a new type of small agricultural tillers or cultivators used by farmers or homeowners. These are also known as power tillers or garden tillers. Compact, powerful and, most importantly, inexpensive, these agricultural rotary tillers are providing alternatives to four-wheel tractors and in the small farmers' fields in developing countries are more economical than four-wheel tractors.

Two-wheel tractor

The higher power "riding" rotavators cross out of the home garden category into farming category, especially in Asia, Africa and South America, capable of preparing 1 hectare of land in 8–10 hours. These are also known as *power tillers* or *walking tractors*. Years ago they were considered only useful for rice growing areas, where they were fitted with steel cage-wheels for traction, but now the same are being used in both wetland and dryland farming all over the world. They have multiple functions with related tools for dryland or paddys, pumping, transportation, threshing, ditching, spraying pesticide. They can be used on hills, mountains, in greenhouses and orchards. Diesel designs are more popular in developing countries than gasoline.

Harrow Tool

A spring-tooth drag harrow.

In agriculture, a harrow (often called a set of harrows in a plurale tantum sense) is an implement for breaking up and smoothing out the surface of the soil. In this way it is distinct in its effect from the plough, which is used for deeper tillage. Harrowing is often carried out on fields to follow the rough finish left by plowing operations. The purpose of this harrowing is generally to break up clods (lumps of soil) and to provide a finer finish, a good tilth or soil structure that is suitable for seedbed use. Coarser harrowing may also be used to remove weeds and to cover seed after sowing. Harrows differ from cultivators in that they disturb the whole surface of the soil, such as to prepare a seedbed, instead of disturbing only narrow trails that skirt crop rows (to kill weeds).

Disc harrows.

There are four general types of harrows: disc harrows, tine harrows (including spring-tooth harrows, drag harrows, and spike harrows), chain harrows, and chain-disk harrows. Harrows were originally drawn by draft animals, such as horses, mules, or oxen, or in some times and places by manual labourers. In modern practice they are almost always tractor-mounted implements, either trailed after the tractor by a drawbar or mounted on the three-point hitch.

Crumbler roller, commonly used to compact soil after it has been loosened by a harrow.

Clydesdale horses pulling spike harrows, Murrurundi, NSW, Australia.

A modern development of the traditional harrow is the rotary power harrow, often just called a power harrow.

Types

In cooler climates the most common types are the *disc harrow*, the *chain harrow*, the *tine harrow* or *spike harrow* and the *spring tine harrow*. Chain harrows are often used for lighter work such as levelling the tilth or covering seed, while disc harrows are typically used for heavy work, such as following ploughing to break up the sod. In addition, there are various types of *power harrow*, in which the cultivators are power-driven from the tractor rather than depending on its forward motion.

Tine harrows are used to refine seed-bed condition before planting, to remove small weeds in growing crops and to loosen the inter-row soils to allow for water to soak into the subsoil. The fourth is a chain disk harrow. Disk attached to chains are pulled at an angle over the ground. These harrows move rapidly across the surface. The chain and disk rotate to stay clean while breaking up the top surface to about 1 inch (3 cm) deep. A smooth seedbed is prepared for planting with one pass.

Harrowing with tractor and disk harrow in the 1940s.

Chain harrowing can be used on pasture land to spread out dung, and to break up dead material (*thatch*) in the sward, and similarly in sports-ground maintenance a light chain harrowing is often used to level off the ground after heavy use, to remove and smooth out boot marks and indentations. Used on tilled land in combination with the other two types, chain harrowing rolls remaining larger soil clumps to the surface where weather breaks them down and prevents interference with seed germination.

All four harrow types can be used in one pass to prepare soil for seeding. It is also common to use any combination of two harrows for a variety of tilling processes. Where harrowing provides a very fine tilth, or the soil is very light so that it might easily be wind-blown, a roller is often added as the last of the set.

Harrows may be of several types and weights, depending on their purpose. They almost always consist of a rigid frame that holds discs, teeth, linked chains, or other means of moving soil—but tine and chain harrows are often only supported by a rigid towing-bar at the front of the set.

In the southern hemisphere, so-called *giant discs* are a specialised kind of disc harrows that can stand in for a plough in rough country where a mouldboard plough cannot handle tree-stumps and rocks, and a disc-plough is too slow (because of its limited number of discs). Giant scalloped-edged discs operate in a set, or frame, that is often weighted with concrete or steel blocks to improve penetration of the cutting edges. This sort of cultivation is usually followed by broadcast fertilisation and seeding, rather than drilled or row seeding.

A drag is a heavy harrow.

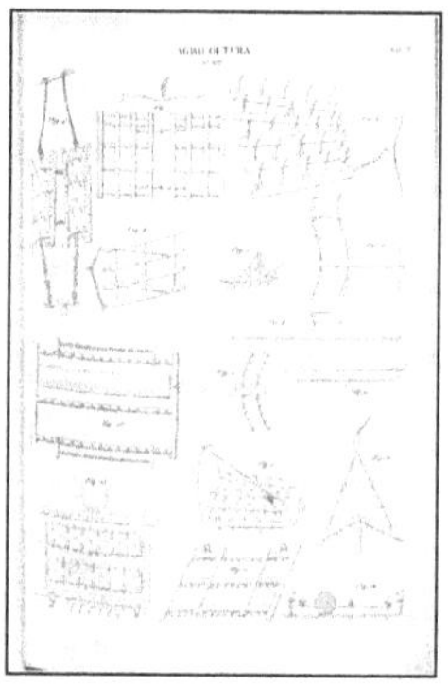

19th century harrow.

Power Harrow

A rotary power harrow, or simply power harrow, has multiple sets of vertical tines. Each set of tines is rotated on a vertical axis and tills the soil horizontally. The result is that, unlike a rotary tiller, soil layers are not turned over or inverted, which is useful in preventing dormant weed seeds from being brought to the surface, and there is no horizontal slicing of the subsurface soil that can lead to hardpan formation.

Historical Reference

An Arabic reference to harrows is to be found in Abu Bakr Ibn Wahshiyya's Nabatean

Agriculture (Kitab al-Filaha al-Nabatiyya), of the 10th century, but claiming knowledge from Babylonian sources. In Europe, harrows were first used in the Middle Ages.

The oldest known illustration of a harrow is in Scene 10 of the eleventh-century Bayeux Tapestry.

Spike harrow depicted on a 16th-century German coat-of-arms.

Plough

Traditional ploughing: a farmer works the land with horses and plough.

A plough (UK) or plow tool or farm implement used in farming for initial cultivation of soil in preparation for sowing seed or planting to loosen or turn the soil. Ploughs are traditionally drawn by working animals such as horses or cattle, but in modern times may be drawn by tractors. A plough may be made of wood, iron, or steel frame with an attached blade or stick used to cut the earth. It has been a basic instrument for most of recorded history, although written references to the plough do not appear in English until c. 1100 at which point it is referenced frequently. The plough represents one of the major agricultural inventions in human history.

The primary purpose of ploughing is to turn over the upper layer of the soil, bringing fresh nutrients to the surface, while burying weeds and the remains of previous crops and allowing them to break down. As the plough is drawn through the soil it creates long trenches of fertile soil called furrows. In modern use, a ploughed field is typically left to dry out, and is then harrowed before planting. Plowing and cultivating a soil homogenizes and modifies the upper 12 to 25 cm of the soil to form a plow layer. In many soils, the majority of fine plant feeder roots can be found in the topsoil or plow layer.

Modern tractor ploughing in South Africa. This plough has five non-reversible mouldboards. The fifth, empty furrow on the left may be filled by the first furrow of the next pass.

Ploughing with oxen. A miniature from an early-sixteenth-century manuscript of the Middle English poem *God Spede þe Plough*, held at the British Museum.

Oxen driven ploughing.

Ploughs were initially human powered, but the process became considerably more efficient once animals were pressed into service. The first animal powered ploughs were undoubtedly pulled by oxen, and later in many areas by horses (generally draft horses) and mules, although various other animals have been used for this purpose. In industrialised countries, the first mechanical means of pulling a plough were steam-powered (ploughing engines or steam tractors), but these were gradually superseded by internal-combustion-powered tractors.

Modern competitions take place for ploughing enthusiasts like the National Ploughing Championships in Ireland. Use of the plough has decreased in some areas, often those significantly threatened by soil damage and erosion, in favour of shallower ploughing and other less invasive conservation tillage techniques.

Natural farming methods are emerging that do not involve any ploughing at all, unless an initial ploughing is necessary to break up hardpan on a new plot to be cultivated, so that the newly introduced soil life can penetrate and develop more quickly and deeply. By not ploughing, beneficial fungi and microbial life can develop that will eventually bring air into the soil, retain water and build up nutrients. A healthy soil full of active fungi and microbial life, combined with a diverse crop (making use of companion planting), suppresses weeds and pests naturally and retains rainwater. Thus the intensive use of water-, oil- and energy hungry irrigation, fertilizers and herbicides are avoided. Cultivated land becomes more fertile and productive over time, while tilled land tends to go down in productivity over time due to erosion and the removal of nutrients with every harvest. Proponents of permaculture claim that it is the only way of farming that can be maintained when fossil fuel runs out. On the other hand, the advantage of agricultural methods that require repeated ploughing are that they allow monocropping on a large scale at remote locations, using industrial machinery rather than human labor.

Etymology

Ploughing.

In older English, as in other Germanic languages, the plough was traditionally known by other names, e.g. Old English *sulh*, Old High German *medela, geiza, huohilī(n)*, Old Norse *arðr* (Swedish *årder*), and Gothic *hōha*, all presumably referring to the ard (scratch plough). The term plough or plow, as used today, was not common until 1700.

The modern word *plough* comes from Old Norse *plógr*, and therefore Germanic, but it appears relatively late (it is not attested in Gothic), and is thought to be a loanword from one of the north Italic languages. Words with the same root appeared with related meanings: in Raetic *plaumorati* "wheeled heavy plough" (Pliny, *Nat. Hist.* 18, 172), and in Latin *plaustrum* "farm cart", *plōstrum, plōstellum* "cart", and *plōxenum, plōximum* "cart box". The word must have originally referred to the wheeled heavy plough, which was common in Roman northwestern Europe by the A.D. 5th century.

Orel (2003) tentatively attaches *plough* to a PIE stem **blōkó-*, which gave Armenian *pe□em* "to dig" and Welsh *bwlch* "crack", though the word may not be of Indo-European origin.

Parts

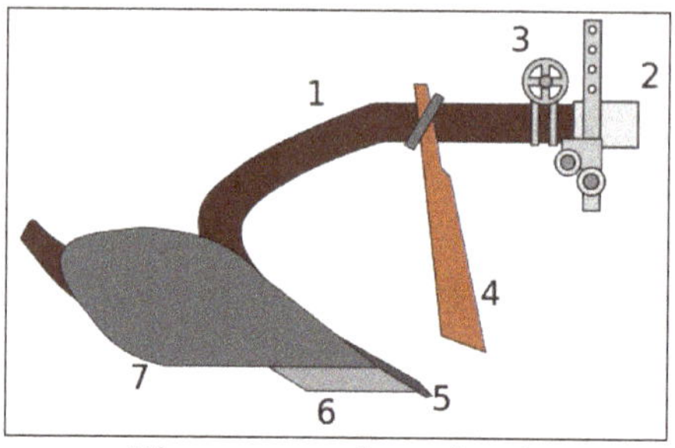

Diagram - modern plough.

The diagram (*right*) shows the basic parts of the modern plough:

1. Beam,
2. Hitch (Brit: hake),
3. Vertical regulator,
4. Coulter (knife coulter pictured, but disk coulter common),
5. Chisel (foreshare),
6. Share (mainshare),
7. Moldboard.

Other parts not shown or labelled include the frog (or frame), runner, landside, shin, trashboard, and stilts (handles).

On modern ploughs and some older ploughs, the mouldboard is separate from the share and runner, so these parts can be replaced without replacing the mouldboard. Abrasion eventually destroys all parts of a plough that come into contact with the soil.

History

Ploughing with buffalo in Hubei, China.

Hoeing

When agriculture was first developed, simple hand-held digging sticks and hoes were used in highly fertile areas, such as the banks of the Nile where the annual flood rejuvenates the soil, to create drills (furrows) to plant seeds in. Digging sticks, hoes, and mattocks were not invented in any one place, and hoe-cultivation must have been common everywhere agriculture was practiced. Hoe-farming is the traditional tillage method in tropical or sub-tropical regions, which are characterized by stony soils, steep slope gradients, predominant root crops, and coarse grains grown at wide distances apart. While hoe-agriculture is best suited to these regions, it is used in some fashion everywhere. Instead of hoeing, some cultures use pigs to trample the soil and grub the earth.

Ard

Ancient Egyptian ard, c.1200 BC. (Burial chamber of Sennedjem).

Some ancient hoes, like the Egyptian *mr*, were pointed and strong enough to clear rocky soil and make seed drills, which is why they are called *hand-ards*. However, the domestication of oxen in Mesopotamia and the Indus valley civilization, perhaps as early as the 6th millennium BC, provided mankind with the draft power necessary to develop the larger, animal-drawn true ard (or scratch plough). The earliest was the *bow ard*, which consists of a *draft-pole* (or *beam*) pierced by a thinner vertical pointed stick called the *head* (or *body*), with one end being the *stilt* (handle) and the other a *share* (cutting blade) that was dragged through the topsoil to cut a shallow furrow ideal for most cereal crops. The ard does not clear new land well, so hoes or mattocks must be used to pull up grass and undergrowth, and a hand-held, coulter-like *ristle* could be used to cut deeper furrows ahead of the share. Because the ard leaves a strip of undisturbed earth between the furrows, the fields are often cross-ploughed lengthwise and widthwise, and this tends to form squarish fields (Celtic fields). The ard is best suited to loamy or sandy soils that are naturally fertilized by annual flooding, as in the Nile Delta and Fertile Crescent, and to a lesser extent any other cereal-growing region with light or thin soil. By the late Iron Age, ards in Europe were commonly fitted with coulters.

Mouldboard Plough

Water buffalo used for ploughing in Si Phan Don, Laos.

To grow crops regularly in less fertile areas, the soil must be turned to bring nutrients to the surface. A major advance for this type of farming was the turnplough, also known as the mouldboard plough (UK), moldboard plow (US), or frame-plough. A *coulter* (or *skeith*) could be added to cut vertically into the ground just ahead of the *share* (in front of the *frog*), a wedge-shaped cutting edge at the bottom front of the *mouldboard* with the landside of the frame supporting the undershare (below-ground component).

The upper parts of the frame carry (from the front) the coupling for the motive power (horses), the coulter and the landside frame. Depending on the size of the implement, and the number of furrows it is designed to plough at one time, a forecarriage with a wheel or wheels (known as a furrow wheel and support wheel) may be added to support the frame (wheeled plough). In the case of a single-furrow plough there is only one wheel at the front and handles at the rear for the ploughman to steer and manoeuvre it.

When dragged through a field the coulter cuts down into the soil and the share cuts horizontally from the previous furrow to the vertical cut. This releases a rectangular strip of sod that is then lifted by the share and carried by the mouldboard up and over, so that the strip of sod (slice of the topsoil) that is being cut lifts and rolls over as the plough moves forward, dropping back to the ground upside down into the furrow and onto the turned soil from the previous run down the field. Each gap in the ground where the soil has been lifted and moved across (usually to the right) is called a *furrow*. The sod that has been lifted from it rests at about a 45 degree angle in the next-door furrow and lies up the back of the sod from the previous run.

In this way, a series of ploughing runs down a field leaves a row of sods that lie partly in the furrows and partly on the ground lifted earlier. Visually, across the rows, there is the land (unploughed part) on the left, a furrow (half the width of the removed strip of soil) and the removed strip almost upside-down lying on about half of the previous strip of inverted soil, and so on across the field. Each layer of soil and the gutter it came from forms the classic furrow.

The mouldboard plough greatly reduced the amount of time needed to prepare a field, and as a consequence, allowed a farmer to work a larger area of land. In addition, the resulting pattern of low (under the mouldboard) and high (beside it) ridges in the soil forms water channels, allowing the soil to drain. In areas where snow buildup causes difficulties, this lets farmers plant the soil earlier, as the snow runoff drains away more quickly.

A reconstruction of a mouldboard plough.

There are five major parts of a mouldboard plough:

1. Mouldboard,
2. Share,
3. Landside,
4. Frog,
5. Tailpiece.

A *runner* extending from behind the share to the rear of the plough controls the direction of the plough, because it is held against the bottom land-side corner of the new furrow being formed. The holding force is the weight of the sod, as it is raised and rotated, on the curved surface of the mouldboard. Because of this runner, the mouldboard plough is harder to turn around than the scratch plough, and its introduction brought about a change in the shape of fields — from mostly square fields into longer rectangular "strips" (hence the introduction of the furlong).

An advance on the basic design was the *iron ploughshare*, a replaceable horizontal cutting surface mounted on the tip of the share. The earliest ploughs with a detachable and replaceable share date from around 1000 BC in the Ancient Near East, and the earliest iron ploughshares from ca. 500 BC in China. Early mouldboards were basically wedges that sat inside the cut formed by the coulter, turning over the soil to the side. The ploughshare spread the cut horizontally below the surface, so when the mouldboard lifted it, a wider area of soil was turned over. Mouldboards are known in Britain from the late 6th century on.

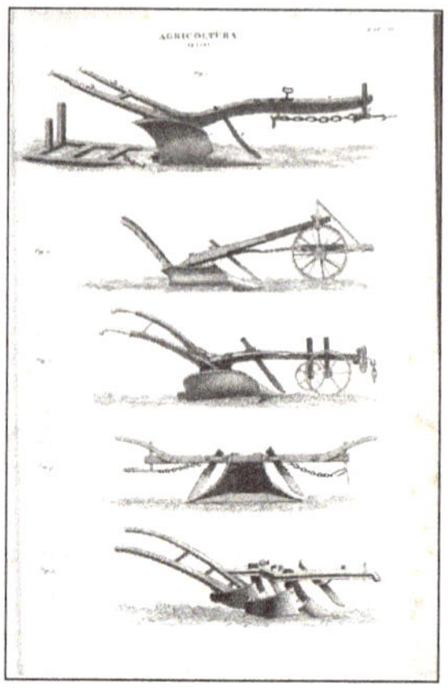

19th century editing plough.

Loy Ploughing

Loy ploughing was a form of manual ploughing in Ireland, on very small farms — or on very hilly ground, where horses could not work or where farmers could not afford them. It was used up until the 1960s in poorer land. This suited the moist climate of Ireland, as the trenches formed by turning in the sods providing drainage. It also allowed the growing of potatoes in bogs as well as on mountain slopes where no other cultivation could take place.

Heavy Ploughs

Chinese iron plough with curved mouldboard, 1637.

In the basic mouldboard plough the depth of the cut is adjusted by lifting against the runner in the furrow, which limited the weight of the plough to what the ploughman could easily lift. This limited the construction to a small amount of wood (although metal edges were possible). These ploughs were fairly fragile, and were not suitable for breaking up the heavier soils of northern Europe. The introduction of wheels to replace the runner allowed the weight of the plough to increase, and in turn allowed the use of a much larger mouldboard faced in metal. These *heavy ploughs* led to greater food production and eventually a significant population increase around AD 600.

Before the Han Dynasty (202 BC –AD 220), Chinese ploughs were made almost entirely of wood, except the iron blade of the ploughshare. By the Han period, the entire ploughshare was made of cast iron; these are the first known heavy mouldboard iron ploughs.

The Romans achieved the heavy wheeled mouldboard plough in the late 3rd and 4th century AD, when archaeological evidence appears, inter alia, in Roman Britain. The first indisputable appearance after the Roman period is from 643, in a northern Italian document. Old words connected with the heavy plough and its use appear in Slavic, suggesting possible early use in this region. The general adoption of the carruca heavy plough in Europe appears to have accompanied the adoption of the three-field system in the later eighth and early ninth centuries, leading to an improvement of the agricultural productivity per unit of land in northern Europe. This was accompanied by larger fields as well, known variously as carucates, ploughlands, and ploughgates.

Improved Designs

'A Champion ploughman', from Australia, circa 1900.

The basic plough, with coulter, ploughshare and mouldboard remained in use for a millennium. Major changes in design did not become common until the Age of Enlightenment, when there was rapid progress in design. Joseph Foljambe in Rotherham, England, in 1730 used new shapes as the basis for the Rotherham plough, which also covered the mouldboard with iron. Unlike the heavy plough, the Rotherham (or Rotherham swing) plough consisted entirely of the coulter, mouldboard and handles. It was much lighter than conventional designs and became very popular in England. It may have been the first plough to be widely built in factories and the first to be commercially successful.

In 1789 Robert Ransome, an iron founder in Ipswich started casting ploughshares in a disused malting at St Margaret's Ditches. As a result of a mishap in his foundry, a broken mould caused molten metal to come into contact with cold metal, making the metal surface extremely hard — chilled casting — which he advertised as "self sharpening" ploughs, and received patents for his discovery.

James Small further advanced the design. Using mathematical methods he experimented with various designs until he arrived at a shape cast from a single piece of iron, an improvement on the *Scots plough* of James Anderson of Hermiston. A single-piece cast iron plough was also developed and patented by Charles Newbold in the United States. This was again improved on by Jethro Wood, a blacksmith of Scipio, New York, who made a three-part Scots Plough that allowed a broken piece to be replaced. In 1837 John Deere introduced the first steel plough; it was so much stronger than iron designs that it could work soil in areas of the US that had previously been considered unsuitable for farming.

Improvements on this followed developments in metallurgy: steel coulters and shares with softer iron mouldboards to prevent breakage, the chilled plough (an early example of surface-hardened steel), and eventually the face of the mouldboard grew strong enough to dispense with the coulter.

Single-sided Ploughing

Single-sided ploughing in a ploughing match.

The first mouldboard ploughs could only turn the soil over in one direction (conventionally always to the right), as dictated by the shape of the mouldboard, and so the field had to be ploughed in long strips, or *lands*. The plough was usually worked clockwise around each land, ploughing the long sides and being dragged across the short sides without ploughing. The length of the strip was limited by the distance oxen (or later horses) could comfortably work without a rest, and their width by the distance the plough could conveniently be dragged. These distances determined the traditional size of the strips: a furlong, (or "furrow's length", 220 yards (200 m)) by a chain (22 yards (20 m)) — an area of one acre (about 0.4 hectares); this is the origin of the acre. The one-sided action gradually moved soil from the sides to the centre line of the strip. If the strip was in the same place each year, the soil built up into a ridge, creating the ridge and furrow topography still seen in some ancient fields.

Turnwrest Plough

The turnwrest plough allows ploughing to be done to either side. The mouldboard is removable, turning to the right for one furrow, then being moved to the other side of the plough to turn to the left (the coulter and ploughshare are fixed). In this way adjacent furrows can be ploughed in opposite directions, allowing ploughing to proceed continuously along the field and thus avoiding the ridge and furrow topography.

Reversible Plough

A four-furrow reversible Kverneland plough.

The reversible plough has two mouldboard ploughs mounted back-to-back, one turning to the right, the other to the left. While one is working the land, the other is carried upside-down in the air. At the end of each row, the paired ploughs are turned over, so the other can be used. This returns along the next furrow, again working the field in a consistent direction.

Riding and Multiple-furrow Ploughs

Early tractor-drawn two-furrow plough.

Early steel ploughs, like those for thousands of years prior, were *walking ploughs*, directed by the ploughman holding onto handles on either side of the plough. The steel ploughs were so much easier to draw through the soil that the constant adjustments of the blade to react to roots or clods was no longer necessary, as the plough could easily cut through them. Consequently, it was not long after that the first *riding ploughs* appeared. On these, wheels kept the plough at an adjustable level above the ground, while the ploughman sat on a seat; whereas, with earlier plows the plowman would have had to walk. Direction was now controlled mostly through the draught team, with levers allowing fine adjustments. This led very quickly to riding ploughs with multiple mouldboards, dramatically increasing ploughing performance.

A single draught horse can normally pull a single-furrow plough in clean light soil, but in heavier soils two horses are needed, one walking on the land and one in the furrow. For ploughs with two or more furrows more than two horses are needed and, usually, one or more horses have to walk on the loose ploughed sod—and that makes hard going for them, and the horse treads the newly ploughed land down. It is usual to rest such horses every half-hour for about ten minutes.

Heavy volcanic loam soils, such as are found in New Zealand, require the use of four heavy draught horses to pull a double-furrow plough. Where paddocks are more square than long-rectangular it is more economical to have horses four wide in harness than two-by-two ahead, thus one horse is always on the ploughed land (the sod). The limits of strength and endurance of horses made greater than two-furrow ploughs uneconomic to use on one farm.

Amish farmers tend to use a team of about seven horses or mules when spring ploughing and as Amish farmers often help each other plough, teams are sometimes changed at noon. Using this method about 10 acres (4.0 ha) can be ploughed per day in light soils and about 2 acres (0.81 ha) in heavy soils.

Improvement in Metallurgy and Design

John Deere, a Vermont (U.S.A.) blacksmith, noted that the ploughing of many sticky, non-sandy soils might benefit from modifications in the design of the mouldboard and in the metals used. He noted that a polished needle would enter leather and fabric with greater ease, and a polished pitchfork required less effort as well. In the pursuit of a polished and thus slicker surface for a plough, he experimented with portions of saw blades and by 1837, he was making polished, cast steel ploughs. The energy effort required was lessened, which enabled the use of larger ploughs, making more effective use of horse power.

Balance Plough

The advent of the mobile steam engine allowed steam power to be applied to ploughing from about 1850. In Europe, soil conditions were often too soft to support the weight of heavy traction engines. Instead, counterbalanced, wheeled ploughs, known as *balance ploughs,* were drawn by cables across the fields by pairs of ploughing engines on opposite field edges, or by a single engine drawing directly towards it at one end and drawing away from it via a pulley at the other end. The balance plough had two sets of ploughs facing each other, arranged so when one was in the ground, the other set was lifted into the air. When pulled in one direction, the trailing ploughs were lowered onto the ground by the tension on the cable. When the plough reached the edge of the field, the other engine pulled the opposite cable, and the plough tilted (balanced), putting the other set of shares into the ground, and the plough worked back across the field.

A German balance plough. The left-turning set of shares have just completed a pass, and the right-turning shares are about to enter the ground to return across the field.

One set of ploughs was right-handed, and the other left-handed, allowing continuous ploughing along the field, as with the turnwrest and reversible ploughs. The man credited with the invention of the ploughing engine and the associated balance plough, in the mid nineteenth century, was John Fowler, an English agricultural engineer and inventor.

In America the firm soil of the Plains allowed direct pulling with steam tractors, such as the big Case, Reeves or Sawyer-Massey breaking engines. Gang ploughs of up to fourteen bottoms were used. Often these big ploughs were used in regiments of engines, so that in a single field there might be ten steam tractors each drawing a plough. In this way hundreds of acres could be turned over in a day. Only steam engines had the power to draw the big units. When internal combustion engines appeared, they had neither the strength nor the ruggedness compared to the big steam tractors. Only by reducing the number of shares could the work be completed.

Stump-jump Plough

Disc ploughs in Australia, circa 1900.

The stump-jump plough was an Australian invention of the 1870s, designed to cope with the breaking up of new farming land, that contains many tree stumps and rocks that would be very expensive to remove. The plough uses a moveable weight to hold the ploughshare in position. When a tree stump or other obstruction such as a rock is encountered, the ploughshare is thrown upwards, clear of the obstacle, to avoid breaking the plough's harness or linkage; ploughing can be continued when the weight is returned to the earth after the obstacle is passed.

A simpler system, developed later, uses a concave disc (or a pair of them) set at a large angle to the direction of progress, that uses the concave shape to hold the disc into the soil — unless something hard strikes the circumference of the disk, causing it to roll up and over the obstruction. As the arrangement is dragged forward, the sharp edge of the disc cuts the soil, and the concave surface of the rotating disc lifts and throws the soil to the side. It doesn't make as good a job as the mouldboard plough (but this is not considered a disadvantage, because it helps fight wind erosion), but it does lift and break up the soil.

Modern Ploughs

A British woman ploughing on a World War I recruitment poster for the Women's Land Army.

Modern ploughs are usually multiple reversible ploughs, mounted on a tractor via a three-point linkage. These commonly have between two and as many as seven mouldboards — and *semi-mounted* ploughs (the lifting of which is supplemented by a wheel about halfway along their length) can have as many as eighteen mouldboards. The hydraulic system of the tractor is used to lift and reverse the implement, as well as to adjust furrow width and depth. The ploughman still has to set the draughting linkage from the tractor so that the plough is carried at the proper angle in the soil. This angle and depth can be controlled automatically by modern tractors. As a complement to the rear plough a two or three mouldboards-plough can be mounted on the front of the tractor if it is equipped with front three-point linkage.

Specialist Ploughs

Chisel Plough

The *chisel plough* is a common tool to get deep tillage (prepared land) with limited soil disruption. The main function of this plough is to loosen and aerate the soils while leaving crop residue at the top of the soil. This plough can be used to reduce the effects of compaction and to help break up ploughpan and hardpan. Unlike many other ploughs the chisel will not invert or turn the soil. This characteristic has made it a useful addition to no-till and low-till farming practices that attempt to maximise the erosion-prevention benefits of keeping organic matter and farming residues present on the soil surface through the year. Because of these attributes, the use of a chisel plough is considered by some to be more sustainable than other types of plough, such as the mouldboard plough.

A modern John Deere 8110 Farm Tractor using a chisel plough. The ploughing tines are at the rear; the refuse-cutting coulters at the front.

Bigham Brother Tomato Tiller.

The chisel plough is typically set to run up to a depth of eight to twelve inches (200 to 300 mm). However some models may run much deeper. Each of the individual ploughs, or shanks, are typically set from nine inches (229 mm) to twelve inches (305 mm) apart. Such a plough can encounter significant soil drag, consequently a tractor of sufficient power and good traction is required. When planning to plough with a chisel plough it is important to bear in mind that 10 to 20 horsepower (7.5 to 15 kW) per shank will be required, depending on depth.

Cultivators are often similar in form to chisel ploughs, but their goals are different. Cultivator teeth work near the surface, usually for weed control, whereas chisel plough shanks work deep beneath the surface. Consequently, cultivating also takes much less power per shank than does chisel ploughing.

Ridging Plough

A ridging plough is used for crops, such as potatoes or scallions, which are grown buried in ridges of soil using a technique called *ridging* or *hilling*. A ridging plough has two mouldboards facing away from each other, cutting a deep furrow on each pass, with high ridges either side. The same plough may be used to split the ridges to harvest the crop.

Scottish Hand Plough

This is a variety of ridge plough notable in that the blade points towards the operator. It is used solely by human effort rather than with animal or machine assistance, and is pulled backwards by the operator, requiring great physical effort. It is particularly used for second breaking of ground, and for potato planting. It is found in Shetland, some western crofts and more rarely Central Scotland. The tool is typically found on small holdings too small or poor to merit use of animals.

Mole Plough

The *mole plough* or *subsoiler* allows underdrainage to be installed without trenches, or it breaks up deep impermeable soil layers that impede drainage. It is a very deep plough, with a torpedo-shaped or wedge-shaped tip, and a narrow blade connecting this to the body. When dragged through the ground, it leaves a channel deep under the ground, and this acts as a drain. Modern mole ploughs may also bury a flexible perforated plastic drain pipe as they go, making a more permanent drain — or they may be used to lay pipes for water supply or other purposes. Similar machines, so called pipe-and-cable-laying ploughs, are even used under the sea, for the laying of cables, as well as preparing the earth for side-scan sonar in a process used in oil exploration.

Paraplough

The paraplough or paraplow is a tool for loosening compacted soil layers 12 to 16 inches deep and still maintain high surface residue levels.

Spade Plough

The spade plough is designed to cut the soil and turn it on its side, minimizing the damage to the earthworms, soil microorganism, and fungi. This helps maximize the sustainability and long term fertility of the soils.

Effects of Mouldboard Ploughing

Mouldboard ploughing, in cold and temperate climates, no deeper than 20 cm, aerates the soil by loosening it. It incorporates crop residues, solid manures, limestone and commercial fertilizers along with oxygen. By doing so, it reduces nitrogen losses by denitrification, accelerates mineralization and increases short-term nitrogen availability for transformation of organic matter into humus. It erases wheel tracks and ruts caused by harvesting equipment. It controls many perennial weeds and pushes back the growth of other weeds until the following spring. It accelerates soil warming and water evaporation in spring because of the lesser quantity of residues on the soil surface. It facilitates seeding with a lighter seeder. It controls many enemies of crops (slugs, crane flies, seedcorn maggots-bean seed flies, borers). It increases the number of "soil-eating" earthworms (endogea) but is detrimental to vertical-dwelling earthworms (anecic).

Ploughing leaves very little crop residue on the surface, which otherwise could reduce both wind and water erosion. Over-ploughing can lead to the formation of hardpan. Typically farmers break up hardpan up with a subsoiler, which acts as a long, sharp knife to slice through the hardened layer of soil deep below the surface. Soil erosion due to improper land and plough utilization is possible. Contour ploughing mitigates soil erosion by ploughing across a slope, along elevation lines. Alternatives to ploughing, such as the no till method, have the potential to actually build soil levels and humus. These may be suitable to smaller, more intensively cultivated plots, and to farming on poor, shallow or degraded soils that ploughing would further degrade.

- Plows in art

Back side of a 100 Mark banknote issued 1908.

1975 Italian Lira coin.

Broadcast Spreader

A broadcast seeder, alternately called a broadcast spreader or centrifugal fertilizer spreader(europe), is a farm implement commonly used for spreading seed, lime, fertilizer, sand, ice melt, etc., and is an alternative to drop spreaders/seeders.

Hand-pushed broadcast spreader.

Types

ATV tow spreaders normally have a larger capacity to enable the coverage of larger areas.

The smallest are handheld with a hopper of several liters and which operate via hand cranking. A bit larger are push units with the spinning disk powered by gearing to the wheels. The next size up is designed to be towed behind a garden tractor or ATV. Very similar in size to the tow behind units are broadcast seeders that mount to the three-point hitch of a compact utility tractor, these are ideal for landscape and small property maintenance. Still larger are commercial broadcast seeders/spreaders designed and sized appropriately for agricultural tractors and mount to the tractor's three point hitch. The broadcast seeders that are mounted to a three point hitch are powered by a power take-off (P.T.O.) shaft from the tractor. At the largest size are pull behind or chassis mounted units for agricultural use that can spread widths of up to 90 feet.

How They Work

View of a tractor-operated broadcast seeder moved by three-point hitch and driven by PTO shaft.

The basic operating concept of broadcast spreads is simple. A large material hopper is positioned over a horizontal spinning disk, the disk has a series of 3 or 4 fins attached to it which throw the dropped materials from the hopper out and away from the seeder/spreader. Alternately a pendulum spreading mechanism may be employed, this method is more common in mid-sized commercial spreaders for improved consistency in spreading. The photos clearly show the material hopper. Hoppers are commonly made of plastic, painted steel, or stainless steel. Stainless steel is usually used in large commercial units for strength and because granular fertilizer is often quite corrosive.

Some seeders/spreaders have directional fins to control the direction of the material that is thrown from the spreader. All broadcast spreaders require some form of power to spin the disk. On hand carried units, a hand crank spins gears to turn the disk. On tow behind units, the wheels spin a shaft that turns gears which, in turn, spin the disk. As is partially visible in one of the photos, with tractor mounted units, a mechanical P.T.O. shaft connected to the tractor and controlled by the tractor operator, spins the disk. There are some seeder/spreaders made for garden size tractors that use a 12 volt motor to spin the dispersing disk and yaw.

Planter Farm Implement

A two row planter featuring John Deere "71 Flexi" row units.

Like a grain drill a planter is an agricultural farm implement towed behind a tractor, used for sowing crops through a field. It is connected to the tractor with a draw-bar, or a three-point hitch. Planters lay the seeds down in precise manner along rows. Seeds are distributed through devices called row units. The row units are spaced evenly along the planter. Planters vary greatly in size, from 1 row to 48, with the biggest in the world being the 48-row John Deere DB120. The space between the row units also vary greatly. The most common row spacing in the United States today is 30 inches.

On smaller and older planters, a marker extends out to the side half the width of the planter and creates a line in the field where the tractor should be centered for the next pass. The marker is usually a single disc harrow disc on a rod on each side of the planter. On larger and more modern planters, GPS navigation and auto-steer systems for the tractor are often used, eliminating the need for the marker. Some precision farming equipment such as Case IH AFS uses GPS/RKS and computer-controlled planter to sow seeds to precise position accurate within 2 cm. In an irregularly shaped field, the precision farming equipment will automatically hold the seed release over area already sewn when the tractor has to run overlapping pattern to avoid obstacles such as trees.

John Deere MaxEmerge XP Planter with Case IH AFS precision farming system which auto-steers using GPS.

Older planters commonly have a seed bin for each row and a fertilizer bin for two or more rows. In each seed bin plates are installed with a certain number of teeth and tooth spacing according to the type of seed to be sown and the rate at which the seeds are to be sown. The tooth size (actually the size of the space between the teeth) is just big enough to allow one seed in at a time but not big enough for two. Modern planters often have a large bin for seeds that are distributed to each row known as central commodity systems.

Drive Systems

There are different types of planters available with the main difference being mechanical driven vs hydraulic/electrical driven. In a mechanical drive system the unit works by a small suspended tire being driven by another which is in contact with the ground (driven) tire. As the operator lowers the planter the two tires make contact and the planter is engaged. When the driven wheel begins to turn it then turns a series of gears that determine the population of the seed produced. The gears can be changed by the operator in order to change the planting population. A hydraulic driven system came about to correct the shortfalls of the ground driven system. Hydraulic driven systems allow the operator to change population on the go, as well as allowing the computer controller to follow a prepared prescription for a individual field. The system also allowed for plant populations to be infinite in that mechanical gears systems are limited to set number of population settings and gears available from manufactures. In 2014 John Deere introduced the ExactEmerge row unit which introduced high-speed planting. Precision Planting (later purchased by John Deere from Monsanto) followed suit and released the vDrive system. These system were unique, not that they electrical, but that they allowed an operator to double their speed when planting. Other manufactures had already developed an electrical planter, but lacked these additional improvements. Traditionally, an operator would plant at about 4.5-5.5 mph for optimal performance, however, with the advent of these systems electrical motors match the speed of the tractor and "dead-drop" the seed in the trench using either a belt or brush-belt which cause the forward momentum of the planter to be offset by the rearward momentum of the seed. Older systems would instead drop the seed through a tube after the meter rather than place is in the seed trench directly.

Drip Irrigation

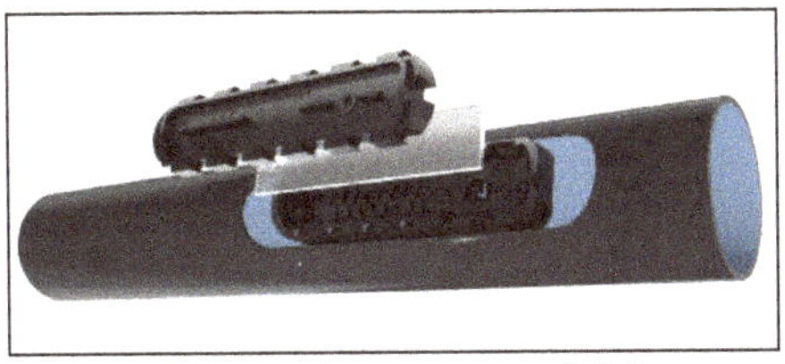

Open pressure compensated dripper.

Drip irrigation is a form of irrigation that saves water and fertilizer by allowing water to drip slowly to the roots of many different plants, either onto the soil surface or directly onto the root zone, through a network of valves, pipes, tubing, and emitters. It is done through narrow tubes that deliver water directly to the base of the plant. It is chosen instead of surface irrigation for various reasons, often including concern about minimizing evaporation.

History

Drip irrigation in Mexico vineyard, 2000.

Primitive drip irrigation has been used since ancient times. Fan Sheng-Chih Shu, written in China during the first century BCE, describes the use of buried, unglazed clay pots filled with water as a means of irrigation. Modern drip irrigation began its development in Germany in 1860 when researchers began experimenting with subsurface irrigation using clay pipe to create combination irrigation and drainage systems. Research was later expanded in the 1920s to include the application of perforated pipe systems. The usage of plastic to hold and distribute water in drip irrigation was later developed in Australia by Hannis Thill.

Usage of a plastic emitter in drip irrigation was developed in Israel by Simcha Blass and his son Yeshayahu. Instead of releasing water through tiny holes easily blocked by tiny particles, water was released through larger and longer passageways by using velocity to slow water inside a plastic emitter. The first experimental system of this type was established in 1959 by Blass who partnered later (1964) with Kibbutz Hatzerim to create an irrigation company called Netafim. Together they developed and patented the first practical surface drip irrigation emitter.

In the United States, the first drip tape, called *Dew Hose*, was developed by Richard Chapin of Chapin Watermatics in the early 1960s.

Modern drip irrigation has arguably become the world's most valued innovation in agriculture since the invention of the impact sprinkler in the 1930s, which offered the first practical alternative to surface irrigation. Drip irrigation may also use devices called micro-spray heads, which spray water in a small area, instead of dripping emitters. These are generally used on tree and vine crops with wider root zones. Subsurface drip irrigation (SDI) uses permanently or temporarily buried dripperline or drip tape located at or below the plant roots. It is becoming popular for row crop irrigation, especially in areas where water supplies are limited or recycled water is used for irrigation. Careful study of all the relevant factors like land topography, soil, water, crop and agro-climatic conditions are needed to determine the most suitable drip irrigation system and components to be used in a specific installation.

Components and Operation

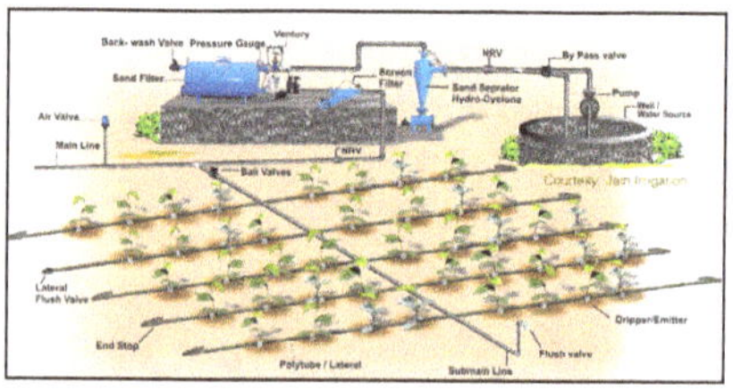

Drip irrigation system layout and its parts.

Water distribution in subsurface drip irrigation.

Nursery flowers watered with drip irrigation in Israel.

Components used in drip irrigation (listed in order from water source) include:

- Pump or pressurized water source,
- Water filter(s) or filtration systems: sand separator, Fertigation systems (Venturi injector) and chemigation equipment (optional),
- Backwash controller (Backflow prevention device),
- Pressure Control Valve (pressure regulator),
- Distribution lines (main larger diameter pipe, maybe secondary smaller, pipe fittings),
- Hand-operated, electronic, or hydraulic control valves and safety valves,
- Smaller diameter polytube (often referred to as "laterals"),
- Poly fittings and accessories (to make connections),
- Emitting devices at plants (emitter or dripper, micro spray head, inline dripper or inline driptube).

In drip irrigation systems, pump and valves may be manually or automatically operated by a controller.

Most large drip irrigation systems employ some type of filter to prevent clogging of the small emitter flow path by small waterborne particles. New technologies are now being offered that minimize clogging. Some residential systems are installed without additional filters since potable water is already filtered at the water treatment plant. Virtually all drip irrigation equipment manufacturers recommend that filters be employed and generally will not honor warranties unless this is done. Last line filters just before the final delivery pipe are strongly recommended in addition to any other filtration system due to fine particle settlement and accidental insertion of particles in the intermediate lines.

Drip and subsurface drip irrigation is used almost exclusively when using recycled municipal waste water. Regulations typically do not permit spraying water through the air that has not been fully treated to potable water standards.

Because of the way the water is applied in a drip system, traditional surface applications of timed-release fertilizer are sometimes ineffective, so drip systems often mix liquid fertilizer with the irrigation water. This is called fertigation; fertigation and chemigation (application of pesticides and other chemicals to periodically clean out the system, such as chlorine or sulfuric acid) use chemical injectors such as diaphragm pumps, piston pumps, or aspirators. The chemicals may be added constantly whenever the system is irrigating or at intervals. Fertilizer savings of up to 95% are being reported from recent university field tests using drip fertigation and slow water delivery as compared to timed-release and irrigation by micro spray heads.

Properly designed, installed, and managed, drip irrigation may help achieve water conservation by reducing evaporation and deep drainage when compared to other types of irrigation such as flood or overhead sprinklers since water can be more precisely applied to the plant roots. In addition, drip can eliminate many diseases that are spread through water contact with the foliage. Finally, in regions where water supplies are severely limited, there may be no actual water savings, but rather simply an increase in production while using the same amount of water as before. In very arid regions or on sandy soils, the preferred method is to apply the irrigation water as slowly as possible.

Pulsed irrigation is sometimes used to decrease the amount of water delivered to the plant at any one time, thus reducing runoff or deep percolation. Pulsed systems are typically expensive and require extensive maintenance. Therefore, the latest efforts by emitter manufacturers are focused toward developing new technologies that deliver irrigation water at ultra-low flow rates, i.e. less than 1.0 liter per hour. Slow and even delivery further improves water use efficiency without incurring the expense and complexity of pulsed delivery equipment.

An emitting pipe is a type of drip irrigation tubing with emitters pre-installed at the factory with specific distance and flow per hour as per crop distance.

An emitter restricts water flow passage through it, thus creating head loss required (to the extent of atmospheric pressure) in order to emit water in the form of droplets. This head loss is achieved by friction / turbulence within the emitter.

Advantages and Disadvantages

Drip irrigation and spare drip irrigation tubes in banana farm at Chinawal, India.

Pot irrigation by On-line drippers.

The advantages of drip irrigation are:

- Fertilizer and nutrient loss is minimized due to localized application and reduced leaching.
- Water application efficiency is high if managed correctly.
- Field levelling is not necessary.
- Fields with irregular shapes are easily accommodated.
- Recycled non-potable water can be safely used.
- Moisture within the root zone can be maintained at field capacity.
- Soil type plays less important role in frequency of irrigation.
- Soil erosion is lessened.
- Weed growth is lessened.
- Water distribution is highly uniform, controlled by output of each nozzle.
- Labour cost is less than other irrigation methods.
- Variation in supply can be regulated by regulating the valves and drippers.
- Fertigation can easily be included with minimal waste of fertilizers.
- Foliage remains dry, reducing the risk of disease.
- Usually operated at lower pressure than other types of pressurised irrigation, reducing energy costs.

The disadvantages of drip irrigation are:

- Initial cost can be more than overhead systems.
- The sun can affect the tubes used for drip irrigation, shortening their usable life.

(This section does not include a discussion of the effects of degrading plastic on the soil content and subsequent effect on food crops. With many types of plastic, when the sun degrades the plastic, causing it to become brittle, the estrogenic chemicals (that is, chemicals replicating female hormones) which would cause the plastic to retain flexibility have been released into the surrounding environment.)

- If the water is not properly filtered and the equipment not properly maintained, it can result in clogging.
- For subsurface drip the irrigator cannot see the water that is applied. This may lead to the farmer either applying too much water (low efficiency) or an insufficient amount of water, this is particularly common for those with less experience with drip irrigation.
- Drip irrigation might be unsatisfactory if herbicides or top dressed fertilizers need sprinkler irrigation for activation.
- Drip tape causes extra cleanup costs after harvest. Users need to plan for drip tape winding, disposal, recycling or reuse.
- Waste of water, time and harvest, if not installed properly. These systems require careful study of all the relevant factors like land topography, soil, water, crop and agro-climatic conditions, and suitability of drip irrigation system and its components.
- In lighter soils subsurface drip may be unable to wet the soil surface for germination. Requires careful consideration of the installation depth.
- Most drip systems are designed for high efficiency, meaning little or no leaching fraction. Without sufficient leaching, salts applied with the irrigation water may build up in the root zone, usually at the edge of the wetting pattern. On the other hand, drip irrigation avoids the high capillary potential of traditional surface-applied irrigation, which can draw salt deposits up from deposits below.
- The PVC pipes often suffer from rodent damage, requiring replacement of the entire tube and increasing expenses.
- Drip irrigation systems cannot be used for damage control by night frosts (like in the case of sprinkler irrigation systems).

Uses

Drip irrigation is used in farms, commercial greenhouses, and residential gardeners. Drip irrigation is adopted extensively in areas of acute water scarcity and especially for crops and trees such as coconuts, containerized landscape trees, grapes, bananas, pandey, eggplant, citrus, strawberries, sugarcane, cotton, maize, and potatoes.

Irrigation dripper.

Drip irrigation for garden available in drip kits are increasingly popular for the home-owner and consist of a timer, hose and emitter. Hoses that are 4 mm in diameter are used to irrigate flower pots.

Grain Drying

Grain drying is process of drying grain to prevent spoilage during storage. The grain drying described in this article is that relating to processes supplementary to the natural ones.

Overview

Continuous mixed flow grain dryer.

Hundreds of millions of tonnes of wheat, corn, soybean, rice and other grains as sorghum, sunflower seeds, rapeseed/canola, barley, oats, etc., are dried in grain dryers. In the main agricultural countries, drying comprises the reduction of moisture from about 17-30%w/w to values between 8 and 15%w/w, depending on the grain. The final moisture content for drying must be adequate for storage. The more oil the grain has, the lower its storage moisture content will be (though its initial moisture for drying will also be lower). Cereals are often dried to 14% w/w, while oilseeds, to 12.5% (soybeans),

8% (sunflower) and 9% (peanuts). Drying is carried out as a requisite for safe storage, in order to inhibit microbial growth. However, low temperatures in storage are also highly recommended to avoid degradative reactions and, especially, the growth of insects and mites. A good maximum storage temperature is about 18 °C.

The largest dryers are normally used "Off-farm", in elevators, and are of the continuous type: Mixed-flow dryers are preferred in Europe, while Cross-flow dryers in the United States. In Argentina, both types are commonly found. Continuous flow dryers may produce up to 100 metric tonnes of dried grain per hour. The depth of grain the air must traverse in continuous dryers range from some 0.15 m in Mixed flow dryers to some 0.30 m in Cross-Flow. Batch dryers are mainly used "On-Farm", particularly in the United States and Europe. They normally consist of a bin, with heated air flowing horizontally from an internal cylinder through an inner perforated metal sheet, then through an annular grain bed, some 0.50 m thick (coaxial with the internal cylinder) in radial direction, and finally across the outer perforated metal sheet, before being discharged to the atmosphere. The usual drying times range from 1 h to 4 h depending on how much water must be removed, type of grain, air temperature and the grain depth. In the United States, continuous counterflow dryers may be found on-farm, adapting a bin to slowly drying grain fed at the top and removed at the bottom of the bin by a sweeping auger.

Grain drying is an active area of manufacturing and research. The performance of a dryer can be simulated with computer programs based on mathematical models that represent the phenomena involved in drying: physics, physical chemistry, thermodynamics and heat and mass transfer. Most recently computer models have been used to predict product quality by achieving a compromise between drying rate, energy consumption, and grain quality. A typical quality parameter in wheat drying is breadmaking quality and germination percentage whose reductions in drying are somewhat related.

Grain Drying Fundamentals

Drying starts at the bottom of the bin, which is the first place air contacts. The dry air is brought up by the fan through a layer of wet grain. Drying happens in a layer of 1 to 2 feet thick, which is called the drying zone. The drying zone moves from the bottom of the bin to the top, and when it reaches the highest layer, the grain is dry. The grain below drying zone is in equilibrium moisture content with drying air, which means it is safe for storage; while the grain above still needs drying. The air is then forced out the bin through exhaust vent.

Allowable Storage Time

Allowable storage time is an estimate of how long the grain needs to be dried before spoilage and maintain grain quality during storage. In grain storage process, fungi or molds are the primary concern. Many other factors, such as insects, rodents, and bacteria, also affect the condition of storage. The lower the grain temperature is, the longer the allowable storage time will be.

Proper Moisture Levels for Safe Storage

It is possible for long period safe storage if grain moisture content is less than 14%, and stored away from insects, rodents and birds. The following figure is the recommended moisture content for safe storage.

Storage duration	Required MC for safe storage	Potential problems
Weeks to a few months storage	14% or less	Molds, discoloration, respiration loss, insect damage, moisture adsorption
Storage for 8 to 12 months	13% or less	Insect damage
Storage of farmer's seeds	12% or less	Loss of germination
Storage for more than 1 year	9% or less	Loss of germination

Equilibrium Moisture Content

Moisture content in grain is related to the relative humidity and the temperature of the surrounding air. Equilibrium moisture content point is the point when grain no longer losing or gaining water when contacting with drying air. The final moisture content of the grain is up to the amount of moisture in the drying air, which is the relative humidity. The low relative humidity means air is dry and it has a large potential of picking up water. The lower the relative humidity is, the drier the air is. In general, one-half reduce in relative humidity is caused by 20° degree increase in air temperature.

Temperature

Heated air may be used in grain drying process. It can not only accelerate moisture migration inside the kernel, but also can evaporate the moisture on the surface. The major problem about heated air for drying process is the kernel temperature. The grain kernel may be damaged by high kernel temperatures. Usually, kernel temperature is lower than the air temperature. For different use of the corn, temperatures vary. For example, for seed corn, the maximum temperature is 110 °F; for livestock feeding corn, the maximum temperature is 180 °F.

Aeration

Aeration process refers to the process of moving air through grain. Airflow is a measurement of the amount of air in cubic feet per minute (CFM). In grain drying process, drying time is largely depended on aeration rates. Without sufficient airflow, grain may be damaged before drying is complete. Fans are used to move air through grain.

Application	Grain aeration rate (cfm/bu)
Quality maintenance	1/50 to 1
Natural-air bin drying	1 to 3
Heated-air bin drying	2 to 12
Batch or continuous-flow column dryers	50 to 150
Fluidization	~400

Drying Cost

The drying cost is made up of two parts: the capital cost and the operating cost. Capital cost is largely depend on the drying rate requirement, and equipment cost. Operating cost refers to fuel, electricity and labor force cost. The amount of energy required to dry a bushel of grain is similar for all the drying methods. Some methods depend largely on natural air, while others may use LP heat or natural gas, which make energy cost vary. Basically, fuel and electrical power are the major portions of the operating cost. Drying cost is based on the B.T.U. consumption of temperature change from environment to desired one.

Classification of Grain Drying Methods

In-storage drying methods

1. Low-temperature drying.
2. Multiple-layer drying.

Batch drying methods

1. Bin batch drying.
2. Column batch drying.

Continuous flow drying methods

1. Cross flow drying.
2. Counter flow drying.
3. Concurrent flow drying.

In-storage Drying Methods

Low-temperature Drying

In-storage drying methods refers to those grain is dried and stored in the same container. Low-temperature drying, also known as near-ambient drying, is one of in-storage drying methods. There are four major factors which influence low temperature drying: the variability of weather, the harvest moisture content, the air flow in the storage bin and the amount of heated air. Most low-temperature dryers are built to dry grain as slowly as possible, while in the same time less spoilage on the grain. It is suggested that low temperature drying system is better operated when the average daily temperature is between 30 °C and 50 °C. Rather than control the drying air temperature, the low-temperature drying focuses on the relative humidity in order to achieve equilibrium moisture content (EMC) in all grain layers. Low-temperature drying process usually takes 5 days to several months depends on several important variables: weather, airflow, initial moisture content and amount of heat used. Among which, airflow is the key factor. Without appropriate airflow rate, spoilage will occur before drying is completed. By using heated air (LP heat, electric heat and solar heat), the relative humidity of the drying air is better controlled to achieve the desired moisture content. Usually, heated air dryer is used when the relative humidity larger than 70%. In electric heat dryers, an electrical resistant heater is usually placed before the fan to heat the airstream. In some case, a humidistat is employed to control the heater. In solar heat dryers, the drying air passes through the solar collector first to be heated (usually 10 to 12 °F rise), then enters the bin through the fan and motor. The advantages of in-storage low temperature drying are quick filling, high quality product, less equipment requirement; while the disadvantages are long drying time, electrical demand if using electric heat, high management skills and uncertain harvest moisture content.

Multiple-layer Drying

Multiple layer drying method refers to the use of LP heat or natural gas in drying corn. Compared to low-temperature methods, multiple-layer drying requires higher temperatures, which results in a shorter allowable storage time. Multiple-layer drying without stirring is the basic multiple-layer drying method, in which airstream is entered through an LP heater by a fan. Usually, the temperature rise after the LP burner is remained low in order to avoid over drying in the bottom layers in the bin. As soon as corn is dried in the bin, the burner is turned off and the fan is used to bring the corn to ambient temperature. The advantages of multiple-layer drying without stirring are little handling of corn, and bin can be used as either dryer or storage; the disadvantages are slow filling and over drying in the bottom layers (Bern and Brumm, 2010). Multiple-layer drying with stirring can not only dry grain equilibrium from top to bottom, but also decrease the air resistance of the grain. Moreover, using stirring system can avoid over drying in bottom layer problem and give a uniform grain moisture content in the whole bin. When drying is complete, the burner is turned off while the fan and stirrer are used to mix the corn to achieve equal moisture content and temperature. The advantages of adding stirring are preventing over drying and accelerating drying and allowable fill rate; the disadvantages of stirring system are additional expenses and decreasing bin capacity.

Batch Drying Methods

Bin-Batch Drying

In batch drying methods, certain amount of grain is placed first, usually 2 to 4 inches, the batch is dried and cooled later, then drying is stopped and batch is removed. The batch dryers are usually operating under this sequence and repeating this sequence for several times. The bin-batch drying methods employ a full perforated floor as the dryer. Without stirring, large variety of equipment is available and the batch can be used as both dryer and cooler, but there may be large moisture gradient from top to bottom and losing time in loading and unloading process. When adding stirring system, unequilibrium moisture content problem is avoided, however, stirrer is an added expense. When using bin-batch roof dryer, time losing problem can be solved. There is a drying floor under the bin roof and the drying fan and burner is installed high on the bin wall. When the drying process is completed, grain is put in the regular bin floor, thus unloading time is reduced. However, there is no wet grain holding in bin-batch roof dryers and there is more expense on machines.

Column Batch Drying

The column formed in this kind of dryer is made up of two vertical perforated steel sheets, which is about 12 inches thick each. The capacity of column batch dryers is too small to store grains. The advantages of column batch, stationary bed dryer are easy to move and the dyer can be used as cooler; while the disadvantages are time losing when cooling, loading and unloading and unequal moisture distribution when drying is completed. When column batch recirculating dryer is used, the moisture content variation problem is avoided, but the additional handling process may result in grain spoilage.

Continuous Flow Drying Methods

Cross Flow Drying

Cross flow dryers is one of the most widely used continuous flow dryers. In the cross flow dryer, the airstream is perpendicular to the grain flow. Then the grain near the drying air is over dried, while on the other side, grain is under dried. Moisture gradient exists when drying is complete. In reality, the lower the airflow rate, the higher the grain moisture content variation between two sides of the column.

Concurrent Flow Drying

In the concurrent flow dryer, both the grain and air are moving in the same direction, which means the wettest grain is subjected to the hottest drying air. The kernels leave the drying region at the same temperature and the same moisture content. Energy efficiency is 40% better compared to cross flow dryer. However, the bed depth must be deeper than 12 inches depth than cross flow type. Thus, fan power requirements are high in this type of dryer.

Counter Flow Drying

In the counter flow dryer, the grain and the air are moving in opposite directions, which mean the driest grain is subjected to the hottest drying air. The kernels leave the drying region at the same temperature and the same moisture as in concurrent flow dryers. The suggesting air temperatures are less than 180 °F because the driest kernels are more likely to be damaged by hot air.

Applications of Grain Drying

Sunflower Drying

For different types of sunflowers, the preservation moisture content is different. Oilseed sunflowers are better dried to 9 percent moisture content, while bird seed sunflowers are 10 percent moisture content. Compared to corn drying, sunflowers are more easily to be dried and kept in safe storage. What's more, high temperature may not have adverse effect on sunflowers kernel, which may be the reason of the fatty acid composition. There was no evidence of damage when air was heated up to 220 °F in drying. However, fine hairs and fibers on the seed coat of sunflowers may cause fire hazard. It is suggested that remove the flaming particles first when heating the sunflowers.

Bean Drying

The seed coat of bean is quite fragile and easy to damaged in cracking and splitting, which may cause loss to the producer. Some studies on beans suggested that in order to avoid cracking, it is better to keep drying air above 40 percent relative humidity.

Corn Drying

When drying corn kernels it is important to keep in mind that cracks and fractures in the corn can lead to many problems in both the storage and processing. The major problem that occurs from high temperature drying and then rapid cooling of the grain is stress-cracking. Stress-cracking is when fractures become present in the corn endosperm. Stress-cracked kernels often absorb water too quickly, are more likely to become broken, and are increasingly susceptible to insect and mold damage during dry storage. In order to reduce the amount of grain that is lost due to stress-cracking, medium temperature and slow cooling, or natural air and low temperature drying methods should be employed.

Mower

A mower is a person or machine that cuts (mows) grass or other plants that grow on the ground. Usually mowing is distinguished from reaping, which uses similar implements, but is the traditional term for harvesting grain crops, e.g. with reapers and combines.

A smaller mower used for lawns and sports grounds (playing fields) is called a *lawn mower* or *grounds mower*, which is often self-powered, or may also be small enough to be pushed by the operator. Grounds mowers have reel or rotary cutters. Larger mowers or *mower-conditioners* are mainly used to cut grass (or other crops) for hay or silage and often place the cut material into rows, which are referred to as *windrows*. *Swathers* (or *windrowers*) are also used to cut grass (and grain crops). Prior to the invention and adoption of mechanized mowers, (and today in places where use of a mower is impractical or uneconomical), grass and grain crops were cut by hand using scythes or sickles.

Mower Configurations

Larger mowers are usually *ganged* (equipped with a number or gang of similar cutting units), so they can adapt individually to ground contours. They may be powered and drawn by a tractor or draft animals. The cutting units can be mounted underneath the tractor between the front and rear wheels, mounted on the back with a three-point hitch or pulled behind the tractor as a trailer. There are also dedicated self-propelled cutting machines, which often have the mower units mounted at the front and sides for easy visibility by the driver. *Boom* or *side-arm* mowers are mounted on long hydraulic arms, similar to a backhoe arm, which allows the tractor to mow steep banks or around objects while remaining on a safer surface.

Mower Types

The cutting mechanism in a mower may be one of several different designs:

Sickle Mower

Eicher tractor with a mid-mounted finger-bar mower.

Sickle mowers, also called *reciprocating mowers, bar mowers, sickle-bar mowers*, or *finger-bar mowers*, have a long (typically six to seven and a half feet) bar on which are mounted fingers with stationary guardplates. In a channel on the bar there is a reciprocating sickle with very sharp sickle sections (triangular blades). The sickle bar is driven back and forth along the channel. The grass, or other plant matter, is cut between the sharp edges of the sickle sections and the finger-plates (this action can be likened to an electric hair clipper).

The bar rides on the ground, supported on a skid at the inner end, and it can be tilted to adjust the height of the cut. A spring-loaded board at the outer end of the bar guides the cut hay away from the uncut hay. The so-formed channel, between cut and uncut material, allows the mower skid to ride in the channel and cut only uncut grass cleanly on the next swath. These were the first successful horse-drawn mowers on farms and the general principles still guide the design of modern mowers.

Rotary Mower

Rotary cutters mounted on a swather.

Rotary mowers, also called *drum mowers*, have a rapidly rotating bar, or disks mounted on a bar, with sharpened edges that cut the crop. When these mowers are tractor-mounted they are easily capable of mowing grass at up to 20 miles per hour (32 km/h) in good conditions. Some models are designed to be mounted in double and triple sets on a tractor, one in the front and one at each side, thus able to cut up to 20 foot (6 metre) swaths.

In rough cutting conditions, the blades attached to the disks are swivelled to absorb blows from obstructions. Mostly these are rear-mounted units and in some countries are called *scrub cutters*. Self-powered mowers of this type are used for rougher grass in gardening and other land maintenance.

Reel Mower

Reel mower.

Reel mowers, also called *cylinder mowers* (familiar as the hand-pushed or self-powered cylinder lawn mower), have a horizontally rotating cylindrical reel composed of helical blades, each of which in turn runs past a horizontal cutter-bar, producing a continuous scissor action. The bar is held at an adjustable level just above the ground and the reel runs at a speed dependent on the forward movement speed of the machine, driven by wheels running on the ground (or in self-powered applications by a motor). The cut grass may be gathered in a collection bin.

This type of mower is used to produce consistently short and even grass on bowling greens, lawns, parks and sports grounds. When pulled by a tractor (or formerly by a horse), these mowers are often ganged into sets of three, five or more, to form a *gang mower*. A well-designed reel mower can cut quite tangled and thick tall grass, but this type works best on fairly short, upright vegetation, as taller vegetation tends to be rolled flat rather than cut.

Home reel mowers have certain benefits over motor-powered mowers as they are quieter and not dependent on any extra form of power besides the person doing the mowing. This is useful not only to lessen dependence on other types of power which may have availability issues, but also lessens the impact on the environment.

Flail Mower

Flail mower.

Flail mowers have a number of small blades on the end of chains attached to a horizontal axis. The cutting is carried out by the ax-like heads striking the grass at speed. These types are used on rough ground, where the blades may frequently be fouled by other objects, or on tougher vegetation than grass, such as brush (scrub). Due to the length of the chains and the higher weight of the blades, they are better at cutting thick brush than other mowers, because of the relatively high inertia of the blades. In some types the cut material may be gathered in a collection bin. As a boom mower (see above), a flail mower may be used in an upright position for trimming the sides of hedges, when it is often called a *hedge-cutter*.

Forestry Mulching

Fecon mulching attachment on a Sennebogen excavator, being used to clear roadside brush in Germany.

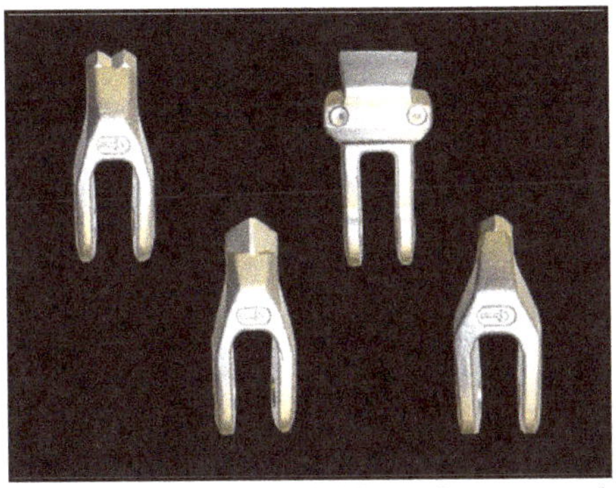

Examples of chipper tools ("teeth") available on forestry mulching attachments.

Forestry mulching is a land clearing method that uses a single machine to cut, grind, and clear vegetation.

A forestry mulching machine, also referred to as a forestry mulcher, masticator, or brushcutter, uses a rotary drum equipped with steel chipper tools ("teeth") to shred vegetation. They are manufactured as application-specific tractors and as mulching attachments ("mulching heads") for existing tracked and rubber-tired forestry tractors, skid steers, or excavators.

Heavy duty forestry mulchers can clear up to fifteen acres of vegetation a day depending on terrain, density, and type of material. Forestry mulchers are often used for land clearing, right-of-way, pipeline/power line, and wildfire prevention and management, vegetation management, invasive species control, and wildlife restoration.

Applications of Forestry Mulching

Right-of-way Clearing and Maintenance

Forestry mulching is used in the right-of-way clearing and maintenance for roads, highways, pipelines, and other utility lines. This process often requires complete removal of standing trees, stumps, and vegetation.

Land Clearing

Forestry mulchers can be used in commercial and residential land clearing projects such as site preparation and development, cutting and clearing brush, nature and recreational trail creation, and seismic exploration.

Wetlands and Riparian Habitat Conservation

Forestry mulching has become popular among nonprofit riparian conservation organizations, government agencies, hunt clubs, and private land owners in attempts to maintain habitats for pheasants, doves, elk, deer and various other animals. Maintaining an animal habitat encompasses several different aspects: food, water, shelter, and space, and there are many products that can help reclaim and maintain wildlife habitats for these animals.

- Food: Forestry mulchers and forestry mowers are often used for removing underbrush and invasive species, such as buckthorn and multiflora rose, in order to allow the rejuvenation of grasses and other food sources.
- Water: Forestry mulchers and tree shears can be used to restore water source access that has been blocked due to tree and understory growth, allowing animals to access the water.
- Shelter and space: Mulching can remove invasive underbrush that prevents the growth of the grasses required by certain animals for shelter, breeding, escaping the summer heat, and protecting themselves from cold temperatures.

Invasive Species Control

Some common invasive plant species such as tamarisk (salt cedar), Pinyon-juniper (pj), Russian olive, buckthorn, and multiflora rose can invade a natural habitat, soak up a tremendous amount of ground water, and need to be removed to reestablish the native habitat or to preserve the water table. Invasive insects such as pine beetles can also devastate forests, leaving behind rotting trees with diminishing timber value and that may become falling hazards if they lose their ability to stand up against wind.

Proactive mulching can reduce stress on trees caused by crowding, making them less susceptible to attack from invasive species. Mulching invasive species in place can control the spread of invasive plants, insects, and fungus. The mulching action tends to discharge the material downward and within a reasonably confined area, versus other methods such as rotary cutters that may laterally disperse pine beetles or other invasive species into neighboring healthy trees.

Wildfire Prevention and Management

- Proactive mulching: Mulching reduces the potential for wildfires by eliminating small leafy plants, fallen or rotten trees, and other fuel sources. If left untreated, these fuel loads increase potential for fire, increase the heat intensity, and serve as fire ladders that enable fire to elevate quickly to the tops of trees which is where a fire can spread most quickly. Mulching can also be used to create a coarse grind finish that can create a more ideal controlled burn.

- Reactive mulching: In addition to proactive thinning of vegetation to mitigate fire fuels, forestry mulching can be used for reactive cutting of lines (fire breaks) on active fires. Larger forestry mulchers leave minimal cleanup requirements and can help reduce the overall costs of active fire mitigation.
- Cleanup: After the fire is out and cleanup efforts are under way, tracked forestry mulching machines, mulching attachments, and an excavator with a mulching attachment can provide a top layer of mulch to prevent soil erosion on slopes and minimize water pollution.

Advantages of Forestry Mulching

By processing trees and other vegetation where they stand, mulching machines eliminate many of the steps involved in land clearing such as site prep, cutting/felling/hauling, and site cleanup. This also eliminates the need for multiple machines such as a bulldozer accompanied by some combination of excavators, tree shears, wood chippers or grinders, and hauling equipment. One simpler jobs only one mulching machine is required, reducing fuel requirements and emissions.

Some mulching machines also have the ability to operate on steep slopes and in small or tight areas, in poor ground conditions, and in wet or snowy weather.

Mulching machines are capable of clearing land of unwanted trees and brush with limited disturbance to soils or desirable vegetation.

Traditional land clearing methods often present an increased risk of erosion by pushing over trees, uprooting the stump and roots, and substantially disturbing soils. In contrast, mulching the vegetation leaves the soil structure intact. The mulched material can be left on the ground and will act as an erosion barrier while returning nutrients back into the soil through decomposition. Over time, grass will naturally grow through the mulch and can be maintained with mowing.

Disadvantages of Forestry Mulching

Depending on the size and orientation of the mulching head, forestry mulchers can only fell smaller trees. Mulchers with a mulching head that rotates about a vertical axis can typically handle trees up to 6-8 inches in diameter, while mulchers with a mulching head that rotates about a horizontal axis can handle trees up to 30 inches in diameter. Mulching trees at the upper end of this size range can be dangerous for both the equipment and the operator. Mulching areas with a variety of vegetation and terrain may require multiple pieces of equipment, including tracked mulching machines, excavator mulchers, and skid steer tractors equipped with mulching attachments.

Rocks and stones cannot be processed or moved by the machines, and the teeth grinding against rocky ground can both wear them out and cause a fire hazard. Smaller rocks and other debris can be thrown through the air, and while these are usually deflected by a protective shroud, they can present a danger to the operator and surrounding people and structures.

Even for the largest machines, mulching is only effective when less than 25 tons of vegetation or 100 trees are present per acre.

Although mulching is significantly faster and less labor-intensive than land clearing by hand, it requires the site to have road access for fueling and maintenance.

Threshing Machine

A threshing machine in operation.

The thrashing machine, or, in modern spelling, threshing machine (or simply thresher), was first invented by Scottish mechanical engineer Andrew Meikle for use in agriculture. It was devised (c. 1786) for the separation of grain from stalks and husks. For thousands of years, grain was separated by hand with flails, and was very laborious and time-consuming, taking about one-quarter of agricultural labor by the 18th century. Mechanization of this process took much of the drudgery out of farm labour.

Early Social Impacts

The Swing Riots in the UK were partly a result of the threshing machine. Following years of war, high taxes and low wages, farm labourers finally revolted in 1830. These farm labourers had faced unemployment for a number of years due to the widespread introduction of the threshing machine and the policy of enclosing fields. No longer were thousands of men needed to tend the crops, a few would suffice. With fewer jobs, lower wages and no prospects of things improving for these workers the threshing machine was the final straw, the machine was to place them on the brink of starvation. The Swing Rioters smashed threshing machines and threatened farmers who had them.

The riots were dealt with very harshly. Nine of the rioters were hanged and a further 450 were transported to Australia.

Later Adoption

Early threshing machines were hand-fed and horse-powered. They were small by today's standards and were about the size of an upright piano. Later machines were steam-powered, driven by a portable engine or traction engine. Isaiah Jennings, a skilled inventor, created a small thresher that doesn't harm the straw in the process. In 1834, John Avery and Hiram Abial Pitts devised significant improvements to a machine that automatically threshes and separates grain from chaff, freeing farmers from a slow and laborious process. Avery and Pitts were granted United States patent #542 on December 29, 1837.

John Ridley, an Australian inventor, also developed a threshing machine in South Australia in 1843.

The 1881 *Household Cyclopedia* said of Meikle's machine:

> "Since the invention of this machine, Mr. Meikle and others have progressively introduced a variety of improvements, all tending to simplify the labour, and to augment the quantity of the work performed. When first erected, though the grain was equally well separated from the straw, yet as the whole of the straw, chaff, and grain, was indiscriminately thrown into a confused heap, the work could only with propriety be considered as half executed. By the addition of rakes, or shakers, and two pairs of fanners, all driven by the same machinery, the different processes of thrashing, shaking, and winnowing are now all at once performed, and the grain immediately prepared for the public market. When it is added, that the quantity of grain gained from the superior powers of the machine is fully equal to a twentieth part of the crop, and that, in some cases, the expense of thrashing and cleaning the grain is considerably less than what was formerly paid for cleaning it alone, the immense saving arising from the invention will at once be seen.
>
> "The expense of horse labour, from the increased value of the animal and the charge of his keeping, being an object of great importance, it is recommended that, upon all sizable farms, that is to say, where two hundred acres [800,000 m^2], or upwards, of grain are sown, the machine should be worked by wind, unless where local circumstances afford the conveniency of water. Where coals are plenty and cheap, steam may be advantageously used for working the machine."

Steam-powered machines used belts connected to a traction engine; often both engine and thresher belonged to a contractor who toured the farms of a district. Steam remained a viable commercial option until the early post-WWII years.

Open-air museum in Saint-Hubert, Belgium.

Farming Process

Threshing is just one step of the process in getting cereals to the grinding mill and customer. The wheat needs to be grown, cut, stooked (shocked, bundled), hauled, threshed, de-chaffed, straw baled, and then the grain hauled to a grain elevator. For many years each of these steps was an individual process, requiring teams of workers and many machines. In the steep hill wheat country of Palouse in the Northwest of the United States, steep ground meant moving machinery around was problematic and prone to rolling. To reduce the amount of work on the sidehills, the idea arose of combining the wheat binder and thresher into one machine, known as a combine harvester. About 1910, horse pulled combines appeared and became a success. Later, gas and diesel engines appeared with other refinements and specifications.

Modern Developments

In Europe and America

Threshing of paddy by machine, Bangladesh.

Modern day combine harvesters (or simply combines) operate on the same principles and use the same components as the original threshing machines built in the 19th century. Combines also perform the reaping operation at the same time. The name *combine* is derived from the fact that the two steps are combined in a single machine. Also, most modern combines are self-powered (usually by a diesel engine) and self-propelled, although tractor powered, pull type combines models were offered by John Deere and Case International into the 1990s.

Today, as in the 19th century, the threshing begins with a cylinder and concave. The cylinder has sharp serrated bars, and rotates at high speed (about 500 RPM), so that the bars beat against the grain. The concave is curved to match the curve of the cylinder, and serves to hold the grain as it is beaten. The beating releases the grain from the straw and chaff.

Whilst the majority of the grain falls through the concave, the straw is carried by a set of "walkers" to the rear of the machine, allowing any grain and chaff still in the straw to fall below. Below the straw walkers, a fan blows a stream of air across the grain, removing dust and fines and blowing them away.

The grain, either coming through the concave or the walkers, meets a set of sieves mounted on an assembly called a shoe, which is shaken mechanically. The top sieve has larger openings, and serves to remove large pieces of chaff from the grain. The lower sieve separates clean grain, which falls through, from incompletely threshed pieces. The incompletely threshed grain is returned to the cylinder by means of a system of conveyors, where the process repeats.

Some threshing machines were equipped with a bagger, which invariably held two bags, one being filled, and the other being replaced with an empty. A worker called a *sewer* removed and replaced the bags, and sewed full bags shut with a needle and thread. Other threshing machines would discharge grain from a conveyor, for bagging by hand. Combines are equipped with a grain tank, which accumulates grain for deposit in a truck or wagon.

A large amount of chaff and straw would accumulate around a threshing machine, and several innovations, such as the air chaffer, were developed to deal with this. Combines generally chop and disperse straw as they move through the field, though the chopping is disabled when the straw is to be baled, and chaff collectors are sometimes used to prevent the dispersal of weed seed throughout a field.

The corn sheller was almost identical in design, with slight modifications to deal with the larger kernel size and presence of cobs. Modern-day combines can be adjusted to work with any grain crop, and many unusual seed crops.

Both the older and modern machines require a good deal of effort to operate. The concave clearance, cylinder speed, fan velocity, sieve sizes, and feeding rate must be adjusted for crop conditions.

Another Development in Asia

Video of a petrol-powered machine threshing rice in Hainan, China.

From the early 20th century, petrol or diesel-powered threshing machines, designed especially to thresh rice, the most important crop in Asia, have been developed along different lines to the modern combine.

Even after the combine was invented and became popular, a new compact-size thresher called a *harvester*, with wheels, still remains in use and at present it is available from a Japanese agricultural manufacturer. The compact-size machine is very convenient to handle in small terrace fields in mountain areas where a large machine, such as combine, is not usable.

People there use this harvester with a modern compact binder.

Preservation

A number of older threshing machines have survived into preservation. They are often to be seen in operation at live steam festivals and traction engine rallies such as the Great Dorset Steam Fair in England, and the Western Minnesota Steam Threshers Reunion in northwest Minnesota.

Musical References

Irish songwriter John Duggan immortalised the threshing machine in a song *The Old Thrashing Mill.* The song has been recorded by Foster and Allen and Brendan Shine.

On the Alan Lomax collection Songs of Seduction (Rounder Select, 2000), there's a bawdy Irish folk song called "The Thrashing Machine" sung by tinker Annie O'Neil, as recorded in the early 20th Century.

In his film score for "Of Mice and Men" (1939) and consequently in his collection "Music for the Movies" (1942), American composer Aaron Copland titled a section of the score "Threshing Machines," to suit a scene in the Lewis Milestone film where Curley is threatening Slim over giving May a puppy, when many of the itinerant worker men are standing around or working on threshers.

In the song Thrasher from the album Rust Never Sleeps, Neil Young compares the modern threshing machine's technique of separating wheat from wheat stalks to the natural forces of time that separate close friends from one another.

Threshing machines appear in Twenty One Pilots' music video for the song House of Gold.

Hog Oiler

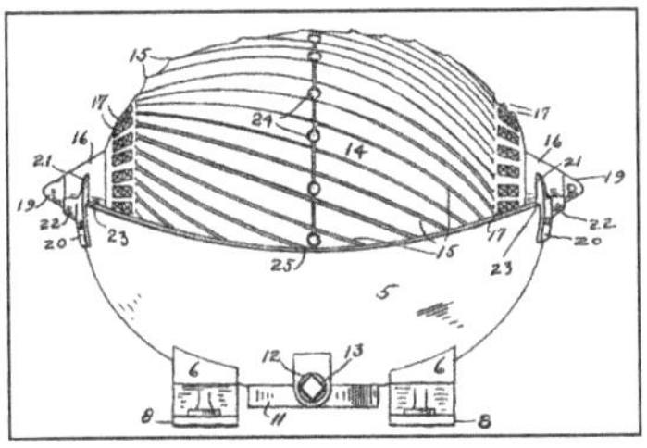

Hog oiler patent.

A hog oiler was a mechanical device employed on farms to be used by hogs to provide relief from insects and offer skin protection. It consisted of a reservoir to hold oil, and a means to distribute the oil onto the hog, often via grooved wheels or cylinders. Hogs seeking relief would rub up against a wheel (or cylinder) causing it to rotate and dispense oil onto their bodies.

History

In the late 1800s and early 1900s hog lice and hog cholera could prove devastating to livestock producers and the hog oiler was seen as one way to combat the problem. The first known patent for a hog oiler device was issued in 1902 by the U.S. Patent Office, however the era of innovation for this device was mainly the years 1913-1923. According to *Goodbye Mr. Louse*, a book by Robert Rauhauser, there may have many as 157 different patented models of hog oilers, but collectors today estimate there could have been as many as 600 manufacturers, most going unpatented.

Prices for hog oilers would range anywhere from four dollars for a small model to twelve dollars and even higher for larger ones with more features. The same companies that manufactured the oilers would often sell special medicated oil to be used with the device, offering further protection. Many farmers however simply chose to use recycled oil or made their own cheaper versions. The U.S. government, while being willing to issue patents, suggested that the oilers might be less than effective, according to collector Bob Coates in *Farm Collector* magazine. "They (the government) recommended mopping or dipping (the hogs) instead" said Coats.

Companies throughout the midwestern U.S., such as Lisle Manufacturing of Clarinda, Iowa, offered farmers and hog producers a variety of styles and sizes including fence-mounted, stand-alone, walk-through and ratchet-governed. However the most common were double-wheel models known as Colubians and Sipes. Often made of cast iron, the early hog oilers could be quite heavy, with some models weighing as much as 150 pounds. Later models from the 1920s onward would mostly be constructed of cheaper, lighter steel and sheet metal. Other known manufacturers of hog oilers included the National Oiler Company of Richmond, Indiana, Rowe Manufacturing, Galesburg, Illinois, and Starbuck Manufacturing, Illinois Implement Company, and O.H.C.Manufacturing, all of Peoria, Illinois.

Collectibles

Antique "watermelon type" hog oiler.

World War II led to a small but growing collectors market for hog oilers today. The cast iron models are considered the most desirable antiques since many originals were gathered up in war scrap iron drives and destroyed. Hog oiler collectors come from all locations and ages, some as young as 11 years old. Generally the "hog belt" of the upper midwest provides the majority of hog oiler finds today—states such as Iowa, Illinois, Indiana, Nebraska and parts of Kansas and Missouri. Prices paid by collectors can vary widely, based on size, functions, and condition of the machine. Smaller hog oilers can be purchased for sometimes thirty dollars or less, while the rarest models can have asking prices in the many thousands of dollars. A small cast iron "wheel" type hog oiler that appeared nearly unused was featured on the History Channel program Pawn Stars on August 8, 2011. It sold for $100.

Baler

A Claas large round baler.

An Abbriata Roto-Baler ejecting a net-wrapped small round bale.

A baler, most often called a hay baler is a piece of farm machinery used to compress a cut and raked crop (such as hay, cotton, flax straw, salt marsh hay, or silage) into compact bales that are easy to handle, transport, and store. Often bales are configured to dry and preserve some intrinsic (e.g. the nutritional) value of the plants bundled. Several different types of balers are commonly used, each producing a different type of bale – rectangular or cylindrical, of various sizes, bound with twine, strapping, netting, or wire.

Industrial balers are also used in material recycling facilities, primarily for baling metal, plastic, or paper for transport.

History

Before the 19th century, hay was cut by hand and most typically stored in haystacks using hay forks to rake and gather the scythed grasses into optimal sized heaps — neither too large (promoting conditions that might create spontaneous combustion, nor too small, so much of the pile is susceptible to rotting. These haystacks lifted most of the plant fibers up off the ground, letting air in and water drain out, so the grasses could dry and cure, to retain nutrition for livestock feed at a later time. In the 1860s mechanical cutting devices were developed; from these came the modern devices including mechanical mowers and balers. In 1872 a reaper that used a knotter device to bundle and bind hay was invented by Charles Withington; this was commercialized in 1874 by Cyrus McCormick. In 1936, Innes invented an automatic baler that tied bales with twine using Appleby-type knotters from a John Deere grain binder; an improved version patented by Ed Nolt in 1939 was more reliable and became commonly used.

Round Baler

The most common type of baler in industrialized countries today is the round baler. It produces cylinder-shaped "round" or "rolled" bales. The design has a "thatched roof" effect that withstands weather well. Grass is rolled up inside the baler using rubberized belts, fixed rollers, or a combination of the two. When the bale reaches a predetermined size, either netting or twine is wrapped around it to hold its shape. The back of the baler swings open, and the bale is discharged. The bales are complete at this stage, but they may also be wrapped in plastic sheeting by a bale wrapper, either to keep hay dry when stored outside or convert damp grass into silage. Variable-chamber large round balers typically produce bales from 48 to 72 inches (120 to 180 cm) in diameter and up to 60 inches (150 cm) in width. The bales can weigh anywhere from 1,100 to 2,200 pounds (500 to 1,000 kg), depending upon size, material, and moisture content. Common modern small round balers (also called "mini round balers" or "roto-balers") produce bales 20 to 22 inches (51 to 56 cm) in diameter and 20.5 to 28 inches (52 to 71 cm) in width, generally weighing from 40 to 55 pounds (18 to 25 kg).

Allis Chalmers Rotobaler.

Originally conceived by Ummo Luebben circa 1910, the first round baler did not see production until 1947 when Allis-Chalmers introduced the Roto-Baler. Marketed for the water-shedding and light weight properties of its hay bales, AC had sold nearly 70,000 units by the end of production in 1960. The next major innovation began in 1965 when a graduate student at Iowa State University, Virgil Haverdink, sought out Wesley F. Buchele, a professor of Agricultural Engineering, seeking a research topic for a master thesis. Over the next year Buchele and Haverdink developed a new design for a large round baler, completed and tested in 1966, and thereafter dubbed the Buchele-Haverdink large round baler. The large round bales were about 5 ft in diameter, 7 ft long, and they weighed about 600 lb after they dried—about 5 lb/ft^3. The design was promoted as a "Whale of a Bale" and Iowa State University now explains the innovative design as follows:

Round baler dumping a fresh bale.

"Farmers were saved from the backbreaking chore of slinging hay bales in the 1960s when Iowa State agricultural engineering professor Wesley Buchele and a group of student researchers invented a baler that produced large, round bales that could be moved by tractor. The baler has become the predominant forage-handling machine in the United States."

In the summer of 1969, the Australian Econ Fodder Roller baler came out, a design that made a 300-lb ground-rolled bale. In September of that same year The Hawkbilt Company of Vinton, Iowa, contacted Dr. Buchele about his design, then fabricated a large ground-rolling round baler which baled hay that had been laid out in a windrow, and began manufacturing large round balers in 1970. In 1972, Gary Vermeer of Pella, Iowa, designed and fabricated a round baler after the design of the A-C Roto-Baler, and the Vermeer Company began selling its model 605 - the first modern round baler. The Vermeer design used belts to compact hay into a cylindrical shape as is seen today. In the early 1980s, collaboration between Walterscheid and Vermeer produced the first effective uses of CV joints in balers, and later in other farm machinery. Due to the heavy torque required for such equipment, double Cardan joints are primarily used. Former Walterscheid engineer Martin Brown is credited with "inventing" this use for universal joints.

By 1975, fifteen American and Canadian companies were manufacturing large round balers.

Transport, Handling, and Feeding

Short-haul Transport and on-field Handling

A large round bale.

Due to the ability for round bales to roll away on a slope, they require specific treatment for safe transport and handling. Small round bales can typically be moved by hand or with lower-powered equipment. Large round bales, due to their size and weight (they can weigh a ton or more) require special transport and moving equipment.

The most important tool for large round bale handling is the bale spear or spike, which is usually mounted on the back of a tractor or the front of a skid-steer. It is inserted into the approximate center of the round bale, then lifted and the bale is hauled away. Once at the destination, the bale is set down, and the spear pulled out. Careful placement of the spear in the center is needed or the bale can spin around and touch the ground while in transport, causing a loss of control. When used for wrapped bales that are to be stored further, the spear makes a hole in the wrapping that must be sealed with plastic tape to maintain a hermetic seal.

Alternatively, a grapple fork may be used to lift and transport large round bales. The grapple fork is a hydraulically driven implement attached to the end of a tractor's bucket loader. When the hydraulic cylinder is extended, the fork clamps downward toward the bucket, much like a closing hand. To move a large round bale, the tractor approaches the bale from the side and places the bucket underneath the bale. The fork is then clamped down across the top of the bale, and the bucket is lifted with the bale in tow. Grab hooks installed on the bucket of a tractor are another tool used to handle round bales, and be done by a farmer with welding skills by welding two hooks and a heavy chain to the outside top of a tractor front loader bucket.

Long-haul Transport

The rounded surface of round bales poses a challenge for long-haul, flat-bed transport, as they could roll off of the flat surface if not properly supported. This is particularly the case with large round bales; their size makes them difficult to flip, so it may not be feasible to flip many of them onto the flat surface for transport and then re-position them on the round surface at the destination. One option that works with both large and small round bales is to equip the flat-bed trailer with guard-rails at either end, which prevent bales from rolling either forward or backward. Another solution is the saddle wagon, which has closely spaced rounded saddles or support posts in which round bales sit. The tall sides of each saddle prevent the bales from rolling around while on the wagon, as the bale settles down in between posts. On 3 September 2010, on the A381 in Halwell near Totnes, Devon, UK an early member of British rock group ELO Mike Edwards was killed when his van was crushed by a large round bale. The cellist, 62, died instantly when the 600 kilograms (1,300 lb) bale fell from a tractor on nearby farmland before rolling onto the road and crushing his van.

Feeding

A large round bale can be directly used for feeding animals by placing it in a feeding area, tipping it over, removing the bale wrap, and placing a protective ring (a *ring feeder*) around the outside so that animals don't walk on hay that has been peeled off the outer perimeter of the bale. The round baler's rotational forming and compaction process also enables both large and small round bales to be fed out by unrolling the bale, leaving a continuous flat strip in the field or behind a feeding barrier.

Silage or Haylage

A recent innovation in hay storage has been the development of the silage or haylage bale, which is a high-moisture bale wrapped in plastic film. These are baled much wetter than hay bales, and are usually smaller than hay bales because the greater moisture content makes them heavier and harder to handle. These bales begin to ferment almost immediately, and the metal bale spear stabbed into the core becomes very warm to the touch from the fermentation process.

Silage or haylage bales may be wrapped by placing them on a rotating bale spear mounted on the rear of a tractor. As the bale spins, a layer of plastic cling film is applied to the exterior of the bale. This roll of plastic is mounted in a sliding shuttle on a steel arm and can move parallel to the bale axis, so the operator does not need to hold up the heavy roll of plastic. The plastic layer extends over the ends of the bale to form a ring of plastic approximately 12 inches (30 cm) wide on the ends, with hay exposed in the center.

To stretch the cling-wrap plastic tightly over the bale, the tension is actively adjusted with a knob on the end of the roll, which squeezes the ends of the roll in the shuttle. In this example wrapping video, the operator is attempting to use high tension to get a flat, smooth seal on the right end. However, the tension increases too much and the plastic tears off. The operator recovers by quickly loosening the tension and allows the plastic to feed out halfway around the bale before reapplying the tension to the sheeting.

These bales are placed in a long continuous row, with each wrapped bale pressed firmly against all the other bales in the row before being set down onto the ground. The plastic wrap on the ends of each bale sticks together to seal out air and moisture, protecting the silage from the elements. The end-bales are hand-sealed with strips of cling plastic across the opening.

Large rectangular baler.

The airtight seal between each bale permits the row of round bales to ferment as if they were in a silo bag, but they are easier to handle than a silo bag, as they are more robust and compact. The plastic usage is relatively high, and there is no way to reuse the silage-contaminated plastic sheeting, although it can be recycled or used as a fuel source via incineration. The wrapping cost is approximately US$5 per bale.

An alternative form of wrapping uses the same type of bale placed on a bale wrapper, consisting of pair of rollers on a turntable mounted on the three-point linkage of a tractor. It is then spun about two axes while being wrapped in several layers of cling-wrap plastic film. This covers the ends and sides of the bale in one operation, thus sealing it separately from other bales. The bales are then moved or stacked using a special pincer attachment on the front loader of a tractor, which does not damage the film seal. They can also be moved using a standard bale spike, but this punctures the airtight seal, and the hole in the film must be repaired after each move.

Plastic-wrapped bales must be unwrapped before being fed to livestock to prevent accidental ingestion of the plastic. Like round hay bales, silage bales are usually fed using a *ring feeder*.

Large Rectangular Baler

Another type of baler in common use, in some areas, will produce large rectangular bales, each bound with a half dozen or so strings of twine, which are then knotted. Such bales are highly compacted and generally weigh somewhat more than round bales. The large rectangular bales are several times larger than the similar small bales. In the prairies of Canada, the large rectangular balers are also called "prairie raptors".

Rectangular Bale Handling and Transport

Large rectangular bales in a field, Charente, France. Sizes of stacks of baled hay need carefully managed, as the curing process is exothermic and the built up heat around internal bales can reach ignition temperatures in the right weather history and atmospheric conditions. Building a deep stack either too wide, or too high increases the risk of spontaneous ignition.

Rectangular bales are easier to transport than round bales, since there is little risk of the bale rolling off the back of a flatbed trailer. The rectangular shape also saves space and allows a complete solid slab of hay to be stacked for transport and storage. Most balers allow adjustment of length and it is common to produce bales of twice the width, allowing stacks with brick-like alternating groups overlapping the row below at right angles, creating a strong structure.

They are well-suited for large scale livestock feedlot operations, where many tons of feed are rationed every hour. Most often, they are baled small enough that one person can carry or toss them where needed.

Due to the huge rectangular shape, large spear forks, or squeeze grips are mounted to heavy lifting machinery, such as large fork lifts, tractors equipped with front end loaders, telehandlers, hay squeezes or wheel loaders, to lift these bales.

Small Rectangular Baler

A small square baler.

A type of baler that produces small rectangular (often called "square") bales was once the most prevalent form of baler, but is less common today. It is primarily used on small acreages where large equipment is impractical, and also for the production of hay for small operations, particularly horse owners who may not have access to the specialized feeding machinery used for larger baled. Each bale is about 15 in x 18 in x 40 in (40 x 45 x 100 cm). The bales are usually wrapped with two, but sometimes three, or more strands of knotted twine. The bales are light enough for one person to handle, about 45 to 60 pounds (20 to 27 kg).

To form the bale, the material to be baled, (which is often hay or straw) in the windrow is lifted by tines in the baler's *reel*. This material is then *packed* into the bale chamber, which runs the length of one side of the baler (normally the right hand side when viewed from the front). A combination plunger and knife move back and forth in the front of this chamber, with the knife closing the door into the bale chamber as it moves backwards. The plunger and knife are attached to a heavy asymmetrical flywheel to provide extra force as they pack the bales. A measuring device—normally a spiked wheel that is turned by the emerging bales—measures the amount of material that is being compressed and, at the appropriate length it triggers the *knotters* that wrap the twine around the bale and tie it off. As the next bale is formed the tied one is driven out of the rear of the baling chamber onto the ground or onto a special wagon or collecting sled hooked to the rear of the baler. This process continues as long as there is material to be baled, and twine to tie it with. The bales emerge with four sides. The twine runs, in two parallel loops, around the wider sides. Of the two narrower sides, there is a cut side and a dull side, and when stacked for storage or transport the bales are normally positioned with the cut side facing outwards.

This form of bale is not much used in large-scale commercial agriculture, because of the costs involved in handling many small bales. However, it enjoys some popularity in small-scale, low-mechanization agriculture and horse-keeping. Besides using simpler machinery and being easy to handle, these small bales can also be used for insulation and building materials in straw-bale construction. Square bales may generally weather better than round bales because a more much dense stack can be put up. However, they don't shed water as round bales do. Convenience is also a major factor in farmers deciding to continue putting up square bales, as they make feeding and bedding in confined areas (stables, barns, etc.) much easier.

Many of these older balers are still to be found on farms today, particularly in dry areas, where bales can be left outside for long periods.

The automatic-baler for small square bales took on most of its present form in 1940. It was first manufactured by the *New Holland Ag* and it used a small petrol engine to provide operating power. It is based on a 1937 invention for a twine-tie baler with automatic pickup.

Wire Balers

Stationary baler.

Bales prior to 1937 were manually wire-tied with two baling wires. Even earlier, the baler was a stationary implement, driven with a tractor or stationary engine using a belt on a belt pulley, with the hay being brought to the baler and fed in by hand. Later, balers were made mobile, with a 'pickup' to gather up the hay and feed it into the chamber. These often used air cooled gasoline engines mounted on the baler for power. The biggest change to this type of baler since 1940 is being powered by the tractor through its power take-off (PTO), instead of by a built-in internal combustion engine.

In present-day production, small square balers can be ordered with twine knotters or wire tie knotters.

Not all stationary wire tying balers used 2 wires. It was not uncommon for the larger bale size (usually 17" x 22") machines to use 'boards' that had three slots for wires and hence tied three wires per bale. Most North American manufacturers produced these machines as either regular models or as size options. 'Small square' three wire tying pick-up balers were available from the early 1930s, principally from J. I. Case & Co. and Ann Arbor. These machines were hand tying and hand threading machines. Although New Holland credits itself with inventing the 'successful small square twine tying baler', it produced such machines for the first time in 1940 after acquiring Ed Nolte and his baler. This baler baled successfully from 1937 onwards. Certainly the quality of the New Holland machines, popularised twine tying hay balers. In Europe, in as early as 1939, both Claas of Germany and Rousseau SA of France had automatic twine tying pick-up balers. Most of these produced low density bales though. The first successful pick-up balers were made by the Ann Arbor Company in 1929. Ann Arbor were acquired by the Oliver Farm Equipment Company in 1943. Despite their head start on the rest of the field, no Ann Arbor balers carried automatic knotters or twisters. Oliver introduced these in 1949.

Square/Wire Bale History

Pickup and Handling Methods

In the 1940s most farmers would bale hay in the field with a small tractor with 20 or less horsepower, and the tied bales would be dropped onto the ground as the baler moved through the field. Another team of workers with horses and a flatbed wagon would come by and use a sharp metal hook to grab the bale and throw it up onto the wagon while an assistant stacks the bale, for transport to the barn.

A later time-saving innovation was to tow the flatbed wagon directly behind the baler, and the bale would be pushed up a ramp to a waiting attendant on the wagon. The attendant hooks the bale off the ramp and stacks it on the wagon, while waiting for the next bale to be produced.

Eventually, as tractor horsepower increased, the thrower-baler became possible, which eliminated the need for someone to stand on the wagon and pick up the finished bales. The first thrower mechanism used two fast-moving friction belts to grab finished bales and throw them at an angle up in the air onto the bale wagon. The bale wagon was modified from a flatbed into a three-sided skeleton frame open at the front, to act as a catcher's net for the thrown bales.

As tractor horsepower further increased, the next innovation of the thrower-baler was the hydraulic tossing baler. This employs a flat pan behind the bale knotter. As bales advance out the back of the baler, they are pushed onto the pan one at a time. When the bale has moved fully onto the pan, the pan suddenly pops up, pushed by a large hydraulic cylinder, and tosses the bale up into the wagon like a catapult.

The pan-thrower method puts much less stress on the bales compared to the belt-thrower. The friction belts of the belt-thrower stress the twine and knots as they grip the bale, and would occasionally cause bales to break apart in the thrower or when the bales landed in the wagon.

Square bale stacker.

Bales may be picked up from the field and stacked using a self-powered machine called a *bale stacker, bale wagon* or *harobed.* There are several designs and sizes. One type picks up square bales are dropped by the baler with the strings facing upward. The stacker will drive up to each bale, pick it up and set it on a three-bale-wide table (the strings are now facing upwards). Once three bales are on the table, the table lifts up and back, causing the three bales to face strings to the side again; this happens three more times until there are 16 bales on the main table. This table will lift like the smaller one, and the bales will be up against a vertical table. The machine will hold 160 bales (ten tiers); usually there will be cross-tiers near the center to keep the stack from swaying or collapsing if any weight is applied to the top of the stack. The full load will be transported to a barn; the whole rear of the stacker will tilt upwards until it is vertical. There will be two pushers that will extend through the machine and hold the bottom of the stack from being pulled out from the stacker while it is driven out of the barn.

A smaller type of stacker.

In Britain (if small square bales are still to be used), they are usually collected as they fall out of the baler in a *bale sledge* dragged behind the baler. This has four channels, controlled by automatic mechanical balances, catches and springs, which sort each bale into its place in a square *eight*. When the sledge is full, a catch is tripped automatically, and a door at the rear opens to leave the eight lying neatly together on the ground. These may be picked up individually and loaded by hand, or they may be picked up all eight together by a *bale grab* on a tractor, a special front loader consisting of many hydraulically powered downward-pointing curved spikes. The square eight will then be stacked, either on a trailer for transport, or in a roughly cubic field stack eight or ten layers high. This cube may then be transported by a large machine attached to the three-point hitch behind a tractor, which clamps the sides of the cube and lifts it bodily.

Storage Methods

Before electrification occurred in rural parts of the United States in the 1940s, some small dairy farms would have tractors but not electric power. Often just one neighbor who could afford a tractor would do all the baling for surrounding farmers still using horses.

To get the bales up into the hayloft, a pulley system ran on a track along the peak of the barn's hayloft. This track also stuck a few feet out the end of the loft, with a large access door under the track. On the bottom of the pulley system was a bale spear, which is pointed on the end and has retractable retention spikes.

A flatbed wagon would pull up next to the barn underneath the end of the track, the spear lowered down to the wagon, and speared into a single bale. The pulley rope would be used to manually lift the bale up until it could enter the mow through the door, then moved along the track into the barn and finally released for manual stacking in tight rows across the floor of the loft. As the stack filled the loft, the bales would be lifted higher and higher with the pulleys until the hay was stacked all the way up to the peak.

When electricity arrived, the bale spear, pulley and track system were replaced by long motorized bale conveyors known as hay elevators. A typical elevator is an open skeletal frame, with a chain that has dull 3-inch (76 mm) spikes every few feet along the chain to grab bales and drag them along. One elevator replaced the spear track and ran the entire length of the peak of the barn. A second elevator was either installed at a 30-degree slope on the side of the barn to lift bales up to the peak elevator, or used dual front-back chains surrounding the bale to lift bales straight up the side of the barn to the peak elevator.

A bale wagon pulled up next to the lifting elevator, and a farm worker placed bales one at a time onto the angled track. Once bales arrived at the peak elevator, adjustable tipping gates along the length of the peak elevator were opened by pulling a cable from the floor of the hayloft, so that bales tipped off the elevator and dropped down to the floor in different areas of the loft. This permitted a single elevator to transport hay to one part of a loft and straw to another part.

This complete hay elevator lifting, transport, and dropping system reduced bale storage labor to a single person, who simply pulls up with a wagon, turns on the elevators and starts placing bales on it, occasionally checking to make sure that bales are falling in the right locations in the loft.

The neat stacking of bales in the loft is often sacrificed for the speed of just letting them fall and roll down the growing pile in the loft, and changing the elevator gates to fill in open areas around the loose pile. But if desired, the loose bale pile dropped by the elevator could be rearranged into orderly rows between wagon loads.

Usage Once in the Barn

The process of retrieving bales from a hayloft has stayed relatively unchanged from the beginning of baling. Typically workers were sent up into the loft, to climb up onto the bale stack, pull bales off the stack, and throw or roll them down the stack to the open floor of the loft. Once the bale is down on the floor, workers climb down the stack, open a cover over a bale chute in the floor of the loft, and push the bales down the chute to the livestock area of the barn.

Most barns were equipped with several chutes along the sides and in the center of the loft floor. This permitted bales to be dropped into the area where they were to be used. Hay bales would be dropped through side chutes, to be broken up and fed to the cattle. Straw bales would be dropped down the center chute, to be distributed as bedding in the livestock standing/resting areas.

Traditionally multiple bales were dropped down to the livestock floor and the twine removed by hand. After drying and being stored under tons of pressure in the haystack, most bales are tightly compacted and need to be torn apart and fluffed up for use.

One recent method of speeding up all this manual bale handling is the bale shredder, which is a large vertical drum with rotary cutting/ripping teeth at the base of the drum. The shredder is placed under the chute and several bales dropped in. A worker then pushes the shredder along the barn aisle as it rips up a bale and spews it out in a continuous fluffy stream of material.

Field of hay bales, lines in field made by baler.

Industrial Balers

A specialized baler designed to compact stretch wrap.

Industrial balers are typically used to compact similar types of waste, such as office paper, Corrugated fiberboard, plastic, foil and cans, for sale to recycling companies. These balers are made of steel with a hydraulic ram to compress the material loaded. Some balers are simple and labor-intensive, but are suitable for smaller volumes. Other balers are very complex and automated, and are used where large quantities of waste are handled.

References

- A.E. Araiza Arizona Daily Star (2010-05-06). "Mechanical grinder quickly creates firebreaks in Oracle". Azstarnet.com. Retrieved 28, June 2020
- R. Goyal, Megh (2012). Management of drip/trickle or micro irrigation. Oakville, CA: Apple Academic Press. p. 104. ISBN 978-1926895123
- U. S. Produce Industry and Labor: Facing the Future in a Global Industry By Linda Calvin retrieved 25, August 2020
- Fox Business News: "Machinery takes the place of migrants as Maine's blueberry harvest booms" September 06, 2015
- Clark, Gregory (2007). A Farewell to Alms: A Brief Economic History of the World. Princeton University Press. p. 286. ISBN 978-0-691-12135-2
- "Tamarac National Wildlife Refuge, Becker County, Minnesota | Woodcock population and young forest habitat management". Timberdoodle.org. Retrieved 28, June 2020
- "Mulchers (Also referred to as Masticators or Brushcutters)". Forest Operations Equipment Catalog. Forests and Rangelands. Retrieved 7, July 2020

Permissions

All chapters in this book are published with permission under the Creative Commons Attribution Share Alike License or equivalent. Every chapter published in this book has been scrutinized by our experts. Their significance has been extensively debated. The topics covered herein carry significant information for a comprehensive understanding. They may even be implemented as practical applications or may be referred to as a beginning point for further studies.

We would like to thank the editorial team for lending their expertise to make the book truly unique. They have played a crucial role in the development of this book. Without their invaluable contributions this book wouldn't have been possible. They have made vital efforts to compile up to date information on the varied aspects of this subject to make this book a valuable addition to the collection of many professionals and students.

This book was conceptualized with the vision of imparting up-to-date and integrated information in this field. To ensure the same, a matchless editorial board was set up. Every individual on the board went through rigorous rounds of assessment to prove their worth. After which they invested a large part of their time researching and compiling the most relevant data for our readers.

The editorial board has been involved in producing this book since its inception. They have spent rigorous hours researching and exploring the diverse topics which have resulted in the successful publishing of this book. They have passed on their knowledge of decades through this book. To expedite this challenging task, the publisher supported the team at every step. A small team of assistant editors was also appointed to further simplify the editing procedure and attain best results for the readers.

Apart from the editorial board, the designing team has also invested a significant amount of their time in understanding the subject and creating the most relevant covers. They scrutinized every image to scout for the most suitable representation of the subject and create an appropriate cover for the book.

The publishing team has been an ardent support to the editorial, designing and production team. Their endless efforts to recruit the best for this project, has resulted in the accomplishment of this book. They are a veteran in the field of academics and their pool of knowledge is as vast as their experience in printing. Their expertise and guidance has proved useful at every step. Their uncompromising quality standards have made this book an exceptional effort. Their encouragement from time to time has been an inspiration for everyone.

The publisher and the editorial board hope that this book will prove to be a valuable piece of knowledge for students, practitioners and scholars across the globe.

Index

www.ingramcontent.com/pod-product-compliance
Lightning Source LLC
LaVergne TN
LVHW080630210926
848485LV00003B/239
9781639890354